SERIES IN THE PHILOSOPHY OF KARL R. POPPER AND CRITICAL RATIONALISM

---

SCHRIFTENREIHE ZUR PHILOSOPHIE KARL R. POPPERS UND DES KRITISCHEN RATIONALISMUS

Herausgegeben von Kurt Salamun

BAND XVI

# BOUNDS OF FREEDOM

## POPPER, LIBERTY AND ECOLOGICAL RATIONALITY

Mahasweta Chaudhury

Rodopi

Amsterdam - New York, NY 2004

For Raj and Mitul

The paper on which this book is printed meets the requirements of 'ISO 9706: 1994, Information and documentation - Paper for documents - Requirements for permanence'.

ISBN: 90-420-1872-0
Editions Rodopi B.V., Amsterdam - New York, NY 2004
Printed in The Netherlands

# CONTENTS

# PREFACE

The present work has its origin in a long-cherished plan to do some evaluation of Feyerabend's iconoclasm about reason and science. Particularly because of its popularity among the postmodernist and feminist critiques of modern science and the universalist concept of rationality.

I realized that we should not underestimate the liberalism of Mill and the critical rationalist account of Popper's somewhat conservative view to which our tradition/generation owes its basic attitude to reason and freedom. As apparent from the title of this work, I have tried here to examine different aspects of the concept of freedom – metaphysical, social and moral – to show that the underlying theme of the different facets is homogeneous. The undercurrent of the theme is individual freedom and open universe. The final output however is a defence of reason and the concept of freedom, which is an essential ingredient of any liberal world-view and democratic political order. In this task the general critical rationalist position is endorsed and amended where this position is not very explicit since it is basically an epistemological theory. The extension of the concept of freedom beyond the human boundary is such an attempt to expand the notion of reason and freedom from a critical rationalist position into a rational ecology.

I have sometimes cited from Indian/Eastern authors and schools of thought to stress some point or given examples from non-western traditions. Though I tried to make it clear, I regret if there is any problem for anyone not familiar with eastern thought or writer. I take this opportunity to acknowledge my debt and express my gratitude to various people and institutions that are instrumental in making this work possible. The first and foremost is the Indian Institute of Advanced Study, Shimla, and its director at that time, Professor Mrinal Miri. The Institute provided me with a fellowship for one year during 1996-97 which relieved me from the regular university job of teaching and grading, and concentrate on research and writing. The serene ambience of the Institute, away from the humdrum of the city life, helped me attain the goal which I could not have achieved in several years in Calcutta. I also thank the excellent staff of the library at the Institute who cooperated with me in every possible way. I am also particularly grateful to Mr A. Jabbar and his staff at the Institute for helping me out with the typing of the first draft of the book. I am also grateful to the University of Calcutta and particularly the Department of Philosophy for granting me extraordinary leave to avail of the fellowship at the Institute. My intellectual debt goes back to Karl Popper and 'fiery' Feyerabend whose tumultuous seminars in the late 'sixties at the London School of Economics initiated me into the thoughts that led to the present work. In this connection I must mention that I owe a lot to professor J. W. N. Watkins of the London School of Economics (who unfortunately died suddenly in 1999) and Professor D. P. Chattopadhyay of Centre for

Studies in Civilization, Delhi who provoked me in various ways through many discussions I had had with them on numerous occasion. My initial disagreements with them (not on the same point) eventually changed some of my previously held ideas. During my stay in Shimla, the late professor Sundar Rajan was also in residence there. He was very keen on my research project and we had many lively discussions. Despite his ill health, he read several parts of my manuscript, made some brilliant remarks and encouraged me to finish the work. Unfortunately he died before the manuscript was ready. I am grateful for many of his suggestions and regret that he did not live to see the book in print. I am grateful to Professor David Miller of Warwick University for suggesting and supplying various materials from time to time particularly the interview of Popper in Italian on pacifism and the green movement. He pointed out various problems which I tried to revise. In this connection I express my gratitude also to Mr.Shyamal K. Ganguly of Calcutta who translated the Italian interview of Popper on the Green Movement into English for me. Finally I thank my husband, Professor Sushil Chaudhury, who despite his busy and tight schedule, spent long hours to help me make the manuscript ready for publication. With my limited knowledge of the computer I could not have done it without his help. The responsibilities for the main work and its lapses however remain solely mine.

Mahasweta Chaudhury
Department of Philosophy
Calcutta University

# INTRODUCTION

## *The Thing Called Freedom*

There are very few concepts like freedom that contain so much multifarious content and controversial element. The term "freedom" also is loaded with theoretical element depending on the conceptual framework or disciplinary matrix if you like.

Leaving aside the semantic jugglery, two broad divisions would be taken up here. One: the political/social freedom with a special reference to the libertarian tradition and its enemies. Two: epistemological indeterminism in general and the extreme position of "no-method " or anarchy. We shall keep the metaphysical aspects of freedom at a distance for our present work except the formal requirement that *freedom* is the condition of being able to choose and to carry out purposes. This formulation has three components:

(1) The primary dictionary meaning, the absence of external restraints; (2) an actual ability with available means and (3) more importantly, an awareness of the power of conscious choice among significant known options. Put it simply, it means that a person is free in so far as she can choose to do something or not, can say "yes" or "no" to a question or command can decide for herself the course of action.

All this may sound hollow because an individual is indeed constrained by physical necessities and conditions most of which are inviolable. Moreover in a civil society there are many social constraints and an increasingly large number of codes one must follow. Stringency of these constraints sometimes makes one feel like a prisoner in the enclosed social space. The complicated tax system, traffic or other civic laws, travel and security measures no doubt practically rob one of much of the acclaimed civil liberty. But these constraints, particularly the latter one (the physical necessities are equally applicable to all) are pre-conditions to maximize similar freedom to the person as well. Civil laws in a more or less free society are enacted to make social life easier with a degree of altruism. So social constraints *per se* are not detrimental to freedom. But only when the basis of the constraint appears to be arbitrary, unnecessary and unreasonable, the freedom of an individual seems to be restricted. It is clear that such unjust restriction can be imposed only by *power*, political power, to be precise. Again if all or most people accept the restrictions, it is difficult to call it restriction of freedom. For example most of us agree to and comply with the restrictions such as tax laws, security and travel regulations, and the like. But certainly democratic system requires a tacit preemptive approval of these restrictions. Other forms of power, particularly rule of one or a few, tends to rob one of this prior approval and leads to infringement of individual right. Of course, in general freedom is now widened as to restrict arbitrary rule and at the same time recognition of rights is extended further by guarantee of law.

Be that as it may, the absence of external restraint has little significance unless there is the positive idea of the requirement of actual ability with available means. Sheer right is an empty concept without appropriate information and opportunity. For example, a beggar or a poor person is *free* to go to a university or a talented village artisan has the *right* to have an exhibition or join an art college. But lack of informational and financial facilities renders that freedom meaningless. So however free the social system is knowledge, education and financial capabilities are necessary to exercise one's right to make a choice.

The most complicated component of social/political freedom is the capacity to make a conscious choice through knowledge and intelligence. In a sense animals and birds are freer than human beings in so far as they pursue their goal instinctively. But that is "blind" groping compared to the conscious and deliberate choice and decisions made by human beings, for the latter has another constraint to overcome to be free, namely *tradition*. The history of five thousand years and so of human civilization has an overwhelming pressure on individuals in the form of complicated customs, creeds, value systems that are positive determinants which influence free decisions. Some people think that these informal external restraints are more powerful than formal (law etc.) restraints and the latter is often based on the former. Tradition in different forms and degrees has sway over an individual choice. The older the civilization, the more rigid and complex its codes are that put a restraint over conscious choice. There are *internal* constraints such as the peculiar sense of dutifulness, reverence, sense of belonging, the feeling of sin or righteousness. *Tradition* and *reason* seem to be two conflicting forces that make it difficult for an individual to make a conscious choice.

Of course a choice made in compliance with the tradition may be rational, but then it does not need a conscious effort to give the reasons in favour of it. But an option *against* the current of tradition needs great efforts to rationalize it in contrast to the option blessed with the approval of the tradition. For their complex and subjective nature, internal constraints are difficult to be identified. To have a more comprehensive concept of freedom someone may prefer Russell's formulation: freedom is 'the absence of obstacles to the realisation of desires'. "Obstacles" obviously include both external and internal restraints. Conformism and consensus are actually the camouflage of lack of freedom. It is difficult to measure the subjective factors of freedom, i.e. *feeling free*. But *being free* can be more clearly indicated and measured by social norms. In a totalitarian rule or even in society of long tradition of rigid conformity, one may feel free with the insular social ambience because one doesn't have informational or actual access to alternative types of society or else the social conditioning is too strong to generate any disapproval. Rabindranath Tagore[1] depicted such well-disciplined conformist society in his play *Tasher Desh* ("The Country of the Cards") where social norms are inflexible and non-conformity or freedom of individual choice is unknown. So the distinction between internal/subjective freedom and external/objective freedom is crucial and the latter only can be judged with more success for its empirical content than the former. Social freedom is external freedom with empirical content and to be reckoned by tangible and reasonably objective norms. Although these two dimensions of freedom are recognized easily and political tyranny can be equally identified as a force against individual freedom, the power of majority rule over an individual or the repressing influence of custom or tradition was often underestimated. The effort of culture, tradition and public opinion over the individual is gradually being recognized. John Stuart Mill's seminal work *On Liberty* (1859) is still the most significant work to indicate that. He notes that there can be a tyranny of the majority (the concept first mentioned by Tocqueville) more enslaving than any tyranny of a political dictator. Majority rule has more than government authority at its disposal, for it can bring to bear upon an individual all the pressures of prevailing opinion and feeling to impose ideas of proper conduct. To defend ourselves against such pressure for conformity, Mill urges that we recognize that 'the sole end for which mankind are warranted, individually or collectively, in interfering with the liberty of action of any of their member is self-protection'[2]. The growing sociological interest and emphasis on the role of authority and tradition tends to breed a sort of conservatism in the recognition and respect for the collective wisdom and experience in their own tradition. One questionable assumption of such a view is that reason itself is immanent in the tradition. Therefore we cannot adjudicate traditions by any transcendent standard. The correct evaluation of a tradition is difficult since there is either a way to regard it as absolutely a force against reason or rather scientific reason. Or on the other hand the sociological approach is to take it as an inevitable process which we cannot come out of.

The pressure of the pre-scientific traditions was at least not that powerful compared to the tradition of science and technology. Herbert Marcuse in his *One Dimensional Man* envisages little freedom in the traditions and organization of modern industrialized societies. He believes that our technologically integrated society has suppressed individuality on behalf of the *efficiency* demanded by the system. The facts that individuals may not feel oppressed and may like the satisfaction of needs which modern technology makes possible only indicate that social forces are moulding and manipulating even the feelings and preferences of individuals. The system place its stamp upon the individuals within it so thoroughly that we are witnessing the creation of one-

---

[1] A Nobel Laureate Bengali poet and great philosopher of the twentieth century.

[2] J. S. Mill, *On Liberty*, p. 10. 6.

dimensional men - creatures incapable of thinking or behaving in any other dimension than the established order.

Instead of elaborating the distinction between what is called negative freedom[3] or absence of constraint (which no one except the proponents of the negative conception of freedom see as enough) and positive freedom which means not only freedom from external constraints but also an ability to rationally order one's desires, we can briefly indicate the difference. Proponents of negative idea of freedom worry by the association of "freedom" with "rational self-mastery" or "selfrealization" since they see the possibility that political or moral authorities may try to "second guess" the purposes and wants of individuals in a way which is paternalistic and inimical to free individual choice. Long before the debate over the positivenegative aspects of freedom, Mill[4] has argued 'that the only purpose for which power can be rightfully exercised over any member of a civilized community, *against his will,* is to prevent harm to others. His own good, either physical or moral, is not a sufficient warrant'. Indeed, he admits that this does not imply that there are no good reasons for remonstrating, reasoning, persuading or entreating someone to follow his long-term interests, or his settled projects, but only that he ought never to be "compelled to do or forbear" because it would make him happier, wiser or more rational.

In similar vein David Thoreau once said if he has the prior information that somebody is coming to him for his (Thoreau's) own good, he will run away from the house. Mills' harm principle and the word "will" have been criticized and subjected to various interpretations, many of which make the concepts either meaningless or vacuous. Gerald Dworkin[5], an advocate of "positive" conception of freedom, has argued that paternalism may be justified, contra-Mill. For example restraint can be exercised upon the individual so that he may live a more rationally ordered life. Dworkin offers a "freedom-maximizing" test for paternalism. That means *one is entitled to intervene coercively only to preserve a wider range of freedom for the individual.* The main difficulty with this principle is that it appears to give too wide a scope for paternalist interference. It may justify for example, not only legislation against motorcyclists without helmet but also for ban on smoking\ drinking or having more than one child.[6] The pursuit of rationality unfortunately may not be valued highly by many individuals. 'Persons sometimes self-consciously choose to nurture an irrational quirk at the centre of their personalities' says Arenson. But the psychological reality of this "irrational quirk"

---

[3] Charles Taylor argues in an article 'What is wrong with negative liberty?' that pretty much everything is wrong with it. See A. Ryan, ed., *The Idea of Freedom*, pp175-193.

[4] J. S. Mill, *On Liberty*, pp. 72-73.

[5] G. Dworkin, *Taking Rights Seriously.*

[6] R. J. Arenson, "Mill vs. Paternalism", in *Ethics*, Vol. 90, No. 4, July 1980, pp. 470-489.

is not uncontroversial and even if true, rationally ordered life would still be regarded as more desirable than unprincipled life or a life (such as one where sanity or sense is diminished) guided by blind force.

We may argue in similar vein to Mill that, in the absence of physical or mental illness, the expressed desires of an individual constitute his "will". However it is also true much of socialist legislation is demonstrably paternalist. Mill's apparent anti-paternalism in *On Liberty* for individual freedom however receives a set back in his programme of science of society which he models in the Newtonian way. It assumes that the laws governing the behaviour of people in social interaction can be inferred mechanically from the laws governing individual people in isolation from society. This thesis closely resembles, some people think,[7] "methodological individualism" proposed by K. Popper.[8] I don't think this is Popper's own view, because he violently rejected Mill's reduction of society into laws of individual atoms. Mill's "individualism" which is "psychologistic" is rejected by Popper by various arguments one of which is that it is the first step towards historicism and also that there is a fundamental difference between the explanation of physical phenomena in the natural science. Popper's tirade against both Mill's psychologism and the historicist tradition indicates that individual freedom can be possible only in an open society, which does not follow any deterministic law. So one might argue that the concept of individual freedom is not logically connected with Mill's variety of social physics. The rational and objective character of scientific laws as upheld by Popper leads to the distinction he made between natural and social phenomena. Again a unity of thought is apparent from the indeterministic and open universe he envisaged extending to the open society and individual freedom he opts for.

Popper's euphoria about the rationality and objectivity of science however is not shared by many others who regard science as a closed tradition which looks more for conformity than criticism. Feyerabend's anti-science diatribe aims to reduce the institution of science to just another tradition unduly respected for its overblown selfimage. Scientific reason is no better than ordinary reason. In a free society the tyranny of science and its "expert knowledge" should be cut to size and pursued by the dictates of common sense reason. The thrust of Feyerabend's view is that a society under the sway of "Western Rationalism", that is of scientism, is more inimical to freedom than the rationalists recognize. He does not say that a free society will have no enforced standards. He says that the standards which won't be *epistemological* will be democratically arrived at, and not imposed by one tradition on everyone.

The apprehension of tyranny of scientific knowledge over other forms is a possible threat to epistemological freedom. The sacrosanct image of science

---

[7] Alan Ryan in *J. S. Mill* expresses such view.

[8] Karl Popper, *The Poverty of Historicism*, p. 136.

and technology, and its sweeping sway over individual is seen by many others to encroach upon the territory of human freedom. Feyerabend's cry for freedom is anarchic. But his rejection of reason (freedom from reason) is not so iconoclastic as it appears to be. Firstly, this rejection is not absolute rejection of reason but only certain *kind of reason*. Second, many others also see the immanent threat in the allpervasive image of science. Herbert Marcuse, Mannheim, Ellul are some examples. Finally, Popper's objective and rational knowledge is not entirely different from the free society Feyerabend opts for in pursuance of scientific activities. The open society and the open universe where indeterminism reigns supreme are not far from Feyerabend's view of "science in a free society". Moreover Popper himself challenged some assumptions of rationalist tradition. These are: uncritical acceptance of metaphysical determinism and reliance on the method of "observationalism".[9] The first one leads to a suspicion that the acceptance of indeterminism may lead to the doctrine of Free Will .The second one is a deep -rooted prejudice and a serious impediment to understand the way scientific method works. Freedom does not require farewell to reason as Feyerabend presumes but only a system that ensures freedom from unreasoned conformity both to epistemological and social norms.

The concept of human freedom is tied up with the notion of non-violence, for any amount of coercion amounts to some form of violence. The assumption of violence being a form of coercion against human being is unwarranted. The reason to relate violence only to human beings and their properties exploits the standard appeal to "respect for others", 'human rights', etc. This is an anthropocentric approach. The arguments for human freedom can be extended not only to other living creatures but also to defend the basic freedom of the rest of nature. I shall argue that this extension of the concept of freedom beyond humans only can establish a rational ecological argument and is compatible with the critical rationalist position.

The cry for freedom dominates the libertarian tradition in social and political philosophy initiated by Mill. Popper strives for both social freedom and epistemological freedom. Although in the latter he recognises rationality, he denies any final principle or truth in science. Feyerabend apparently denies reason and truth, both in society and in science. Or at least he claims that he does.

An interesting move has been made by John Watkins to search for an adequate view of freedom from the critical rationalist position.[10] First thing to revise in this account is 'Kant's uniform appraisal of all people from saints to

---

[9] Karl Popper, "Towards A Rational Theory of Tradition", in *Conjectures and Refutations,* pp. 122-123.

[10] John Watkins, *Human Freedom after Darwin, A Critical Rationalist View,* Open Court, 1999, chapter 8, 'A Third View of Freedom'.

hardened scoundrels as equally free or equally unfree.'[11] Watkins rightly thinks that a view of freedom should be more discriminatory to make a distinction among various degrees of freedom instead of treating it as an all-or-nothing affair. It would be simplistic to divide those who are free and those who are unfree. The important thing is not classifying people in this respect but evaluating different human conditions regarding freedom. The classical notion of freedom takes the notion of self-determination rather abstractly. No individual person can be thought of completely independent of others or in entire independence of everything external to her. This bold step takes us further in the analysis of freedom that allows degrees of autonomy as well as kinds of freedom that we spoke of earlier in the discussion.

The three freedom scales[12] proposed by Watkins are actually the degrees in which people can overcome or distance themselves from some kind of heteronomy. They are as follows: 1) extreme internal heteronomy of the unsuspecting mind by some external agent like super-hypnosis; 2) extreme external heteronomy where a subject's bodily behaviour is subject to completely under the control of some outside agent. Watkins added a third to these two kinds of determinations. He calls it: 3) extreme *disonomy* in tune with *autonomy* and *heteronomy*. This is a case where selfcontrol is lost without passing to an external agent. The three states are cases of "absolute zeros" in the freedom scale but there are no upper limits of freedom. These states have lowering effects on autonomy in different degrees.

The problem of human autonomy was also a major theme in Kant's conception of enlightenment and the role of reason. The enlightenment ideal which is endorsed by the critical rationalist position on freedom and individual autonomy is enshrined in all democratic constitutions. Thus it can be a strong position to thwart all kinds of tyranny and orthodoxy. This conception of freedom and autonomy will be philosophically adequate as Watkins claimed. I think it shall be politically comprehensive as well to combat all kinds of fundamentalism that is threatening the world today. For the emphasis on reason and autonomy alone can stop the slide into irrationalism of various sorts encouraged by anti-science critiques of different stripes. But before going into that question, let us first start with what can be called Mill's social physics which seems to be a reductionist programme - a subversion of individualism. Ibid., p. 222.

[11] Ibid., p. 221.
[12] Ibid., p. 222.

# *CHAPTER I*

## MILL'S SOCIAL PHYSICS AND INDIVIDUAL LIBERTY: THE PROGRAMME AND ITS PROBLEMS

Mill has an ambitious project of having a science of society modelled after physics, particularly Newtonian mechanics; one can of course see the main aim of such a project. If some laws can explain social phenomena laws, then it will lead to successful prediction and ultimately control (far from the libertarian ideal!) of social change.

Of course the difference between social phenomena and physical phenomena is apparent, and Mill realizes this. Accordingly he modified[1] the deductive simplicity of physics and opted for a "method of inverse deduction"[2]. It is apparent that the laws governing the behaviour of people in social interaction can be inferred from the laws governing individual people in isolation from society. Alan Ryan likens this thesis to Popper's "methodological individualism". But that is not correct as we will see later. Mill's methodological reductionism in connection with social behaviour involves a fallacy even on the psychological level. A mass behaviour (such as a protest march) or what is called crowd psychology/anger or violent mood is not just the sum-total of each individual's behaviour taken out of the context of a social level. It is a new emergent phenomenon that cannot be explained in terms of single individual's action in isolation of a society. What an individual could not have done on her own can do that when in league with others. So the *explanandum* (any mass action) cannot be deduced from the *explanans,* (single individual actions) isolated from the social level.[3]

There are two assumptions for Mill's belief that there can be a science of society similar in methodological structure to physics. The first one is already mentioned above, namely the laws governing social behaviour can be derived from the laws governing atoms of individual experience. It is not apparent what these "agents", "data", "elements", are supposed to be. Alan Ryan thinks they are human feelings, desires, aversions,[4] etc. These elements conflict and react with one another and by the composition of forces modify themselves. But we see in reality that the situations are much more complex than that and any calculation or prediction of actual phenomena is difficult. For example, in a situation of ordinary complexity if elements x and y both desire d, it is not predictable

---

[1] Alan Ryan, *John Stuart Mill*, Chap. IX.

[2] J. S. Mill, *A System of Logic*, VI, ix, I.

[3] Alan Ryan, *Mill,* p.149.

[4] Ibid., p. 151.

how they would behave; 1) x may grab it; 2) y may grab it; 3) they may both forego and d goes to a third z or 4) none of x and y modifies his desire and the sheer violence destroy their relationship as well as the object d.

The second assumption is Mill's belief in the uniformity of scientific method. This assumption is backed by the consideration that historical explanation does not involve deductive inference of *explanandum* from *explanans*. But that again is based on a dubious premise that rationality of a method does not require backing all explanations with general laws. We shall come back to this point later in connection with Popper and his review of what he calls Mill's "psychologism". Mill starts with generalization about psychological phenomena, the elementary laws of mind. But he never describes the difficulty of predicting human behaviour in terms of its complex and indeterministic nature but ascribes the difficulty to the 'extraordinary number and variety of the data or elements of the agents, which in obedience to that small number of laws co-operate towards the effect'[5]. Mill finds the problem only to be empirical, not of a methodological nature. It only shows, in his view, the limited "compass of human faculties". But he does not think that this difficulty makes social science unobtainable because we can still check the result of our deductions, and the confidence in that inference is drawn not from a priori reasoning but from 'the consilience between its results and those of observation a posteriori'[6]. But this cannot be the solution as Ryan pointed out because the problem of checking deductions once they are made is not the real difficulty; the real problem lies in making the deductions in the first place.

It seems that Mill is not unaware of the problem because although originally the methodology of social science was modelled after physics he realizes that it cannot after all "be a science of positive predictions, but of tendencies"[7]. That means it is not possible to predict how people *will* behave in certain circumstances only, how they *would behave* in the absence of unforeseen modifying conditions. For example, it is impossible to predict how people would behave if there is an epidemic or an earthquake but it is possible to foresee their behaviour if certain factors were absent e.g. if there were no organized relief measures, medical equipments or ready state actions or if they never faced such situation before. Mill admits that sheer knowledge of *tendencies* (based on the knowledge of human nature) is inadequate for prediction,[8] but it is sufficient for explanation of the past events. But again too many tendencies complicate the scene so much that the original difficulty of inference remains.

---

[5] J.S. Mill, *System of Logic,* VI, ix.
[6] Ibid.
[7] Ibid.
[8] Ibid .VI, IX, 2.

Mill recognizes this difficulty for most social phenomena except the area of economics where the social facts are different from other types of social facts. And tendencies found in one social condition cannot remain unchanged in another social situation, or also in the same society under changed conditions. Mill himself admits this but at the same time finds some types of social facts which are different from others in being *less* influenced by the *consensus* of social phenomena. Economics as pointed out before is such a science. But here again the laws of economics are undoubtedly more stable than other types of social laws but to work as an adequate *explanans*, lot of initial conditions are necessary which are inter-linked with social facts of other types. So prediction here is also difficult because in principle the possibility of an unknown variable is always there. And an adequate explanation can only go from the effect to cause but not from cause (*explanans*) to effect (*explanandum*). So a statement about a social fact can never be explained *before* it has happened but only *after* its actual occurrence.

It seems that Mill realizes the difficulty of his original programme and opts for middle level generalizations instead of general laws. Following Comte this option gives away a lot of Mill's ambitious project and makes a big concession to the empiricists. The fundamental idea of what he calls "inverse deduction" is that where 'it is not possible to deduce truths about the actual course of history from the *law of human nature'*, we may nonetheless be able to guess at "axiomata media", middle level propositions which we verify, not by testing against the facts of history but by seeing whether they are desirable from the laws of human nature or whether at any rate they are compatible with those laws.[9] Ryan has argued that if he 'opened the gate a little, Mill now leaves it to swing wide'. It is true (what Mill said earlier) that experiments in social matters are not possible and no observation can in principle be ever adequate for the conclusions of the inductivist canons.

Now methodologically it is an extreme position but one may nevertheless argue that even if it is not a fully justified conclusion, history can at least show us similar trends or empirical laws which obtain in several societies in reasonably similar conditions. For example, in most colonial societies there were freedom movements or mass movements against the ruling power after a certain period or colonial rule of exploitation and injustice of various sort. These empirical laws (or trends) only indicate some "middle-level, derivative" laws governing different successive states of society. These derivative laws are verified by deduction from the laws of human nature. One may argue that by these derivative laws it is possible for Mill to infer from the principle of human nature via "axiomata media" *to trends of social phenomena.* But Mill's claim is just to tie up the derivative laws with laws of human nature, 'naturally to be expected as the conse-

---

[9] A. Ryan, *Mill,* p. 154.

quences of those ultimate ones'.[10] It seems from the expression that the relation is less than a logical one of derivative inference. This means that there has been no question of either verification or falsification by testing against the *laws of human nature.* Mill's belief in a hierarchy of science (mathematical sciences underlying the physical sciences and the latter underlying other scientific investigations) leads to the position that psychology is basic in the sense that laws of physics are basic to astronomy or atomic physics and many other sciences. Moral science, social sciences are not possible, in Mill's view, without human beings and the latter follow laws of psychology. Thus the general science of society is the last stage in the deduction of social science from individual psychology. In Mill's view people's action can be reduced to the play of psychological atoms. We can at least in principle build up constructions in the form of not only persons but also up to the point of a whole society. This atomistic/reductionist programme aims at explaining gross regularities in terms of breaking them up to regularities belonging to components of psychological nature.

The most well known treatment this programme received so far is at the hand of Karl Popper[11]. Popper approves of Mill's anti-holistic view of society as against those who uphold social wholes such as collectives and groups as real. It rightly insists that the "behaviour" and the "actions'' of collectives such as states and social groups must be reduced to the behaviour and to the actions of human individuals.[12] However he quickly adds in the next line: 'But the belief that the choice of such an individualistic method implies the *choice of a psychological method is mistaken.*'[13] It is clear that to Popper it is "naive to assume" that all social laws are derivable (at least in principle) from the laws of 'human nature. The main thesis of psychologism in his words is 'The doctrine that, society being the product of interacting minds, social laws must ultimately be reducible to psychological laws, since the events of social life, including its conventions, must be the outcome of motive springing from the minds of individual men'.[14] Popper offers an argument from the institutionalist point of view that motives alone cannot fully explain an action unless placed in the general situation namely the social environment. So the institutionalist may contend that sociology *cannot* be reduced to psychological analysis. An adequate analysis of a social action requires other social factors. Sociology that is any account of society is *autonomous* to a great extent.

---

[10] J. S. Mill, *System of Logic* VI, X, 4.

[11] Karl Popper, *The Poverty of Historicism* and *Open Society and its Enemies,* II, contain arguments against Mill's 'psychologism'.

[12] Karl Popper, *Open Society* II, p. 91.

[13] Ibid., p. 91, emphasis added.

[14] Ibid., p. 91, emphasis added.

He anticipates the possible reaction of psychologism in this respect. The proponent of psychologism may continue to argue that the difference between natural and social sciences is that the environments in social situations are man-made. Therefore it can be ultimately explained in terms of human nature. To use Popper's example the economic institution of 'market' can be derived in the last analysis from the psychology of "economic man". Mill would analyze it in terms of the psychological phenomena of "the pursuit of wealth"[15] The followers of psychologism may also point out that the peculiar "psychological structure of human nature" makes the institutions play an important role in society. And gradually these become a part of the tradition or more or less stable social environment. The final point that provoked much of Popper's criticism is the decisive point of any psychologistic view that 'The *origin as well as the development* of traditions must be explicable in terms of human nature'[16]. The origin of any tradition, in this view is some human motive or other. In his enthusiasm against any type of holism, Popper endorses Mill's ant-holism to a great extent. Mill's remark: 'Men are not, when brought together, converted into another kind of substance',[17] is not entirely true. As I have pointed out before, men's single behaviour and behaviour in a group is not always the same. On the contrary often it acquires completely different quality. Examples abound. Communal riots, protest movements, strike etc. Surprisingly Popper (despite his apparent anti-psychologism) regards this remark of Mill as "... one of the most praiseworthy aspects of psychologism, namely its sane opposition to collectivism and holism... Psychologism is, I believe correct only in so far as it insists upon what may be called 'methodological individualism' as opposed to 'methodological collectivism'..."[18] The "dangerous ground" Popper finds in psychologism is that it is led to adopt *historicist method.* If we try to reduce the social facts to psychological facts, it leads to issues about origins and development, i.e. eventually to what Popper calls historicist. It is true that Mill realized the difficulty of psychologism, but he regards it as one of empirical nature. Any generalizing in social science is difficult he thinks, because of "tremendous complexity of the interaction of so many individual minds". The logician Mill is cautious to draw a conclusion 'until sufficient grounds can be pointed out in human nature', he admits that it is not possible to determine a priori the order in which human development will take place or 'the general facts of history up to the present time'. His ground seems to be inductive, as it appears from the following; 'after the first few terms of the series, the influence exercised over each generation by the generation which preceded it becomes ... more and more preponderant over

---

[15] J. S. Mill, *System of Logic*, VI, IX.

[16] Ibid. I,VI,VII.

[17] J.S. Mill, *System of Logic*, VI, VII, IX and id. p. 91.

[18] Ibid. p. 91.

all other influences. The social condition so to say becomes an important determinant'.

As Popper points out first, the weakness of psychologism-especially its historicist version, lies in the ultimate reduction of all social institutions into laws of human nature with all its actions and passions. All social regularities then could be given in terms of "historico-causal" development. For Popper collectives do not exist; ontologically there are no such things as states, nations, armies or other such social groups. Logically statements about such 'dubious' entities can all be translated into statement about 'non dubious'' entities, such as individual men and women. Collectives are abstract concepts, models or theoretical constructs to understand the behaviour of the individuals. Popper here almost endorses Mill's kind of individualism. He says: 'Even "the war" or "the army" is *abstract* concepts, strange as this may sound to some. What is concrete are the many who are killed; or the men and women in uniform etc.'[19]

But of course Mill's individualism goes further than that. He reduces even an individual person's action to atoms of psychological events like hope, desires pain etc. If individualism of both Popper's and Mill's variety is pursued to a great length it will face the difficulties like any other reductionist programme, for example a statement about a protest march if translated into statement about individual persons a, b, c, it can never adequately explain the matter. Because unless we add the sense or purpose of a's participation, b's participation and c's participation - those actions separately do not have the same meaning as when done collectively with a purpose. And once we introduce purpose - which is irretrievably linked up with theoretical and sociological elements-reduction to individualism is given up. This much about Mill and Popper's common programme. Undoubtedly both deny the view that social terms are basic. But Popper does not share Mill's "psychologism" and the only Marxist doctrine he endorses is its anti-psychologistic stance. His basic objections to psychologism[20] are threefold. The first is that human nature is greatly influenced by social institutions, so individual behaviour is conditioned by social norms and states to such a great extent that it is difficult to say how much is in its nature'[21] and how much is due to social conditions. The study of human nature presupposes these institutions and not vice versa. The second difficulty he finds is that social science deals with the *unintended* consequences of people's actions, not with the *intended* ones.

---

[19] Karl Popper, *The Poverty of Historicism*, p. 135.

[20] M. Chaudhury, 'Objective Knowledge and psychologism' *Journal of Indian Council of Philosophical* Research, January, 1995 contains an account of information-processing model of knowledge that is 'psychologistic' in Popper's sense but still yields objective knowledge.

[21] Popper's example: fear of snakes is not inherent but inculcated by education. *Open Society* II, p. 90.

These are social facts and not to be explained in terms of psychological laws.[22] For example if everybody desire an object, the price of that object will *go up* although those who want to buy will *naturally* want the price to be down. Popper's example: a prospective buyer in the real estate market will want the price to be low as a buyer but by entering the market (intended action) he raises the price (unintended consequence) level of real estate. And the third and very important objection is that there is a threat of 'historicism' in the interest in human nature, which leads to questions about how society originated. Popper regards Mill's psychologism to lead to a 'desperate position forced upon him'. 'It is a desperate position because his theory of a pre-social human nature which explains the foundation of society - a psychologistic version of the 'social contract' - is not only a *historical myth*, but also, as it were, a *methodological myth.*' He thinks that if a reduction were to be attempted at all it would be better to interpret psychology in terms of sociology rather than 'the other way round'.

Ryan thinks that methodological individualism would be meaningless if it does not allow reduction of social laws to the laws of 'individual behaviour'. That makes sense if we consider Mill's statement that character of men is formed by both the social circumstances and the basic human nature they are endowed with.[23] According to Ryan Popper is unfair to Mill for the only way in which Mill wants to reduce social laws to the laws of individual psychology is that he wants to link the laws of man in society to the laws of basic behaviour of man. Ryan is right about Mill's recognition that there are very few basic human capacities and most of the diversities in their actions are due to the difference in the circumstances. And this recognition is very close to Popper's own view. As regards the question of unintended consequences, we should note that those who would regard all social actions as consequences of human intention will be contenders of the 'conspiracy theory of history'. But Mill is found (as pointed out by Ryan) to have opposed his father's view that politics is a successful conspiracy of sinister plots. Moreover as an economist Mill explains price rise of say scarce goods not as an organized phenomenon but as a natural consequence of efforts which aims at other goals than price rise. Popper also overemphasized the origin of history issue, because it would be necessary for Mill to raise the questions about the origin of society only if he would want to give a causal or deterministic account of social development. But if we allow Mill the account of mutual interaction of human nature and the institutions created by it, his position is not far from that of Popper. Because if the mutual interaction is recognized then it is not possible even in principle to predict the whole of human history from knowledge of human nature and its history. So Ryan's criticism of Popper is basically correct that the latter is mistaken about Mill's so called "psychologism" as purely

---

[22] Karl Popper, ibid., II, pp. 92-93.

[23] J. S Mill, *System of Logic*, VI, x, 3.

reductionist. Mill has definitely shown more nuances than the simplistic reduction of social facts to human intentions as ascribed to him by Popper.

### *Causal Explanation and Rational Explanation*

But the more serious problem is still open to any attempt at giving an explanation of social science especially of history. Surely there is a fundamental difference between the explanation of physical phenomena and that of human behaviour in history. Mill's notion of individual liberty as espoused in *On Liberty* and his account of individual behaviour in society is not consistent. His programme of social physics aims at giving a causal explanation of history in the same way as physics aims to give an explanation of physical events. Historical explanations are *not* causal explanations. They are narratives at best offering some rational explanation. Mill's initial assumption that there is no methodological difference between physics and history is unwarranted. It is important to realize this distinction and see that history *does not* use laws as physics does. This distinction depends again on a further distinction between *rational* and *causal* explanation. Human behaviour can at best be given a rational explanation (one might also question that) but that again only *after* the action has happened. Mill saw that rationality is a requirement for an individual to be socially/politically free, but for social changes he offers only a causal explanation modeled after explanations in physical science.

One might wonder what kind of explanation for social science he has in mind. It cannot be the covering law model[24] even if a set of laws, instead of one as required by the original covering-law model. Example: It can be shown that there is a threat of plunging into the successful explanation only of analytic truth, not concrete social fact. It can also be argued[25] that the use of law to cover an explanation is not wrong for it is difficult to frame laws because of *the non-explanatory nature of laws*. Deducing a social fact from a law does not explain the phenomena. The causal sequence should be analyzed into simpler components in order to throw light on the chain of events. To give an intelligible explanation, a 'causal narrative' is more important than a *logical* relation

Finally the distinction drawn between causal and rational explanation is essential to understand that the latter is nothing but an attempt to justify the actions of historical agents. Dray alleges in this vein that historical explanations are attempts to observe things from the agent's perspective, to justify or rationalize the action. Causal explanation again is analogous to the deduction of the effect

---

[24] W. Dray, Laws *and Explanation in History*, chap. V.

[25] As done by Dray.

from the causal law and the singular statement about the presence of the cause[26]. Mill does not make this distinction between the "inside'' of the agent and the "outside" view of the spectator and thus was led to the weak doctrine that individuals are prone to atomistic analysis of their behaviour. And Popper is right that it belies Mill's assumption that causal deterministic explanation of human behaviour can be given. This seems to be a threat not only to reconcile historical explanation with the same methodology as physics but also in connection with the basic requirement of rationality of belief (which is a pre-condition for individual freedom) and action. The other problem lurking behind any variety of social physics is the one of prediction of future changes in social actions and institutions. Popper finds any such attempt as a deterministic move. No methodology in principle can predict future social phenomena on the basis of historical laws.

## *Trend, Tradition and Social Law*

Mill's belief in the uniformity of rational explanation led to an attempt to establish uniformity of science. He was concerned to explain far-reaching social changes. In that effort he identifies two kinds of social enquiry: (1) which is concerned with the question of what effect will follow from a given cause, a given social circumstance, (2) what laws determine those social circumstances? In other words" ... what are the causes which produce, and the phenomena which characterize, states of society generally".[27] Any science of society includes the study of social uniformities of succession or change. These may be called "social statics and social dynamics". Mill realized that only freedom could not guarantee good social or political order unless same conditions are fulfilled. The conditions are: 'first: There has existed for all who were accounted citizens - for all who were not slaves kept down by brute force - a system of *education,* beginning with infancy and continued through life, of which whatever also it might include, one main and incessant ingredient was *restraining discipline...* The second condition of permanent political society has been found to be the existence, in some form or other, of the feeling of allegiance or loyalty. This feeling may vary in its objects, and is not confined to any particular form of government; but whether in a democracy or in a monarchy, its essence is always the same; viz. that there be in the constitution of the state *something* which is settled, something permanent, and not to be called in question... The Third *essential* condition (of stability in political society)... is a strong and active principle of rationality'[28].

[26] Karl Popper, *Poverty of Historicism,* pp. 147-152.
[27] J. S. Mill, *System of Logic,* VI x, 1.
[28] Ibid., VI, x, 5.

The uniformity that exists among simultaneous phenomena must be then the result of the co-existence of phenomena at some earlier time, in conjunction with the laws, which determine the succession of phenomena. To Mill the principles for social change (the succession of one social state after another) are of two types: cyclical and cumulative or "progressive change." Popper[29] rejects the suggestion of a "trajectory" or cumulative change as it is simplistic. He also finds agreement between the views of Marx and of Mill, which apparently seems very unusual. Popper points out; 'Thus when Marx says in the preface to *Capital*, "It is the ultimate aim of this work to lay bare the... law of motion of modern society" he might be said to carry out Mill's programme'.

Mill distinguishes clearly between two kinds of sociological inquiry - Popper calls the first one social technology and the second one corresponding to historicist prophecy. Popper thinks that Mill characterized the second one as the 'general science of society by which the conclusions of the other and more special kind of inquiry must be limited and controlled'[30]. This general science of society is based upon the principle of causality, and Mill describes this causal analysis of society as the "Historical Method". 'Mill's "states of society" with 'properties ... changeable from age to age' correspond exactly to Marxist "historical periods', and Mill's optimistic belief in progress resembles Marx's although it is of course much more naive than its dialectical counterpart'.[31] The human affairs move in conformity with two possible astronomical motions, "an orbit" or "a trajectory"[32]. Popper ascribes more simplistic account of social change to Mill than to Marxist dialectic which is more complicated like a "corkscrew movement". He finds further similarity between Mill and Marx, like both of their dissatisfaction with laissez faire liberalism and both 'tries to provide better foundations for carrying into practice the fundamental idea of liberty'[33].

But Mill's study of society is different from that of Marx in an important way, while the former takes the analysis of society into laws of human nature, and laws of the mind, the latter takes social laws, particularly economic laws as more fundamental particularly "the progressiveness of the human". Popper as mentioned before calls this approach 'psychologism' which both Marx and Popper (possibly the only point they share) challenged. No social or political structures can be explained by the general 'progressiveness of human nature'. Popper rejects the assumption that human beings *naturally* (?) shows disposition or natural inclination for progress, he rightly points out that it is not impossible

---

29 Karl Popper, *Open Society*, II, p. 88.
30 Ibid. p. 87.
31 Ibid. VI, x. 6; J. S. Mill, *Civilization, Dissertations and Discussions*, Vol. I, p. 163; J. S. Mill, Review of De Tocqueville's Democracy in America; emphasis added.
32 Ibid. p. 88.
33 Ibid.

that by an epidemic or a virus people may someday aspire only for Nirvana[34]. Marx went the other way, the social laws that determine the nature of social events in his view - are ultimately economic laws. Marx's critique of Mill's psychologism at least leaves sociological laws as autonomous that are not to be explained in terms of psychological principles.

Popper's ascription of reductionism to Mill for explaining historical or social change is simplistic. Trajectory or cumulative change as envisaged by Mill (and also Popper) involves more complex elements than just atoms of human factors. Individuals are changed by social situations and the latter undergoes change by individual persons. Indeed Mill speaks of "progressive change", but that is only to take account of the fact that the major social phenomena change from one period of history to another progressively, but at the same time it is recognized that the change occurs '...both in the character of the human race, and in their outward circumstances so far as moulded by themselves'[35]. Mill only tried to ascertain from such progressive change - a trend or tendency (law) of that change. But trend or law here is not taken in any causal or inviolable way. He is clearly making a distinction between causal law and social law (trend). Historical changes are only empirical laws - summative in nature; predictions of future events are not possible on the basis of historical change. The latter is only a pattern, a tendency among particularly facts to understand the particular situation.

It seems that Mill is very close to Popper's distinction between a law and a trend. Social laws are laws in the sense of a trend or a tendency; they cannot have any predictive power unless they depend on some natural laws or uniformities. The laws of social change are thus merely empirical, '...without any means of determining their real limits, and of judging the changes which have hitherto been in progress are destined to continue indefinitely, or to terminate, or even to be reversed'[36]. What is the motive to *know* the prevailing trends of social change?' Is it just a historical quest or to utilize the knowledge of the trends for some other purpose? Mill's reply to these questions would be clearly for some practical purpose namely: the knowledge of the trends would empower us more than it was in the past to have greater control over social change and in changing its pattern. So although Mill criticized the French historians for confusing trends with natural laws (which only have some predictive power) - he is definitely heading towards a determinist view of social change. This is in conflict with his active advocacy for individual liberty in society.

Moreover his skepticism about the predictive power of social trends is only apparent and based on *merely* empirical reasons. The trends of the order and

---

[34] Karl Popper, *Poverty of Historicism*, p. 157.

[35] Mill, *System of Logic*, VI, x, 3.

[36] Ibid., p. 6.

pattern of human development is unpredictable to him not because individuals are basically free agents, but because history up to the present time has a cumulative influence over individuals and their situations in such a way that it is "impossible" to predict their behaviour in a certain situation. Although he is cautious about inference or prediction from the past to the future, the reason for the impossibility is not based on any logical ground. The reason is based on contingent difficulty of computation because the social circumstances change the individuals so much that their acquired characters are very far from the basic human nature and thus are "impossible" to be predicted in a certain situation of history. So Mill found it difficult to establish any relation between human nature and social change because observation of human history and its trend can only give us some empirical law sans any predictive power. He claims nevertheless to have achieved some low-level tendencies in social change. For example, '... as society advances, mental *tends* more and more to prevail over bodily qualities and masses over individuals:'[i] Mill is inspired by Tocqueville for his theory about 'The irresistible tendency of the age toward *equality*'.

Although the statements expressing low-level tendencies cannot guarantee any methodological advantage over empirical generalization, Mill still retains an ambitious programme. He finds at least one element as a great determinant of change in the states of society: the element is the 'states of the speculative faculties of mankind'. Of course he does not say that desire for knowledge and truth in most individuals is universal, in fact it is found to be pretty low compared to the desire for comfort and wealth. But the desire for comfort and wealth can be gratified only if there is adequate technology to provide for it. And technology depends on the state of the science. Thus progress of technology is incumbent on progress of knowledge and science. So even if direct inspection of history *does not* show a trend towards more knowledge and truth, the motivation for wealth and comfort goad on individuals in a society towards more knowledge and prosperity.

This argument even in its most acceptable form cannot establish that analysis of human motivation and the study of trend can *explain* social change. First, the study of human desires can never be *exhaustive*; there are many periods in history, many different kinds of society. So even if there are *trends,* they are so numerous and diverse that it is not easy to find out some common bonds. Second, if human nature changes with changes in social conditions, *trends* also change with that of social conditions, therefore it is again difficult to distinguish between the elements in *trends* that are *natural* and those, which are due to social conditioning. Third, even if all these empirical difficulties are overcome by computational ability, we still face logical problem; the historical evidences however exhaustive and analyzed can at most show that the trends *were* such and such up to a point of time t. It can only say 'so far as the observation goes a *trend is found* to be of such and such nature', and cannot say that 'this trend will continue'. On Mill's own submission social conditions change from one state of society to another. Then the initial conditions of every state are bound to be different, and there is no way to expect any *tendency* to continue. Popper's tirade against any historicist explanation of social change in this connection still remains unmet. As mentioned earlier he visualized that an epidemic or virus can in future wipe the human desire for wealth, and it is quite conceivable that in future there would be a universal desire for Nirvana.

There is another attempt by Mill to provide a principle on which a theory of history can be constructed. He is convinced both by facts of history and by the general principles of human nature that any appreciable change in the social conditions '... has been preceded by a change, of proportional extent, in the state of their knowledge or in the prevalent beliefs'[37]. Apparently this sounds better since science and knowledge is indeed a dominant factor in social change. We shall come back in a later chapter to the status of science and education with reference to Feyerabend's diatribe against it. It is a fact that features of open and closed societies depend very much on the kind of education they propagate. The state policy of education makes different kind of individual and different kind of society in the end. We should recall that Mill was inimical to uniform mass education, which may produce uniform set of beliefs detrimental to individuality[38]. Mill's optimism about the positive achievements of sciences is comparable to that of Popper, but his Newtonian framework is too naive, '... the circle of human knowledge will be complete and it can only thereafter receive further enlargement by perpetual expansion from within'[39]. Apart from the obvious inconsistency of knowledge being 'complete' and receiving 'perpetual expansion', this statement belies a simplistic trust on mechanical principles and their predictive power (about future events). It belies also his view that historical trends are not absolute but only conditional; not capable of prediction[40] but only give us a guidance about the future.

The observation of the trend shall enable us to change its course by different mode of social policy, education to reverse it if it is inimical to individual freedom. If we interpret Mill's social physics in this way, much of Popper's criticism of it would appear as misguided. The three main objections are as follows: one: Popper finds 'holism in it', two: a belief in absolute trends or tendencies which Mill confuses with genuine causal laws and three: a simplistic belief in social progress. As regards the objection regarding holism, that is taking the states of society as a whole, it is true that Mill argues that in social matters there is such an interaction between various aspects that we cannot isolate the clear causal link.

---

[37] J. S. Mill, *System of Logic,* VI, X, 7.

[38] J. S. Mill, *Dissertation and Discussions*, Vol. I.

[39] J. S. Mill, *System of Logic*, VI, X, 8.

[40] Ibid., p. 4.

For example an economic phenomenon cannot be explained only with pure economic causes without reference to other social factors.

Popper made it clear[41] that the holistic approach tends to explain everything by the same theory. The doctrine of holism holds that every aspect of social life is intimately connected with every other. There are two problems related to the doctrine: a) explanation of no phenomenon can be adequate because the relevant causes would be infinite or else there are too many social facts, and to be precise, not only social, but natural phenomena also can be causally linked with social facts, then it would be an impossible task to explain a single phenomenon. b) tendency to explain social phenomena as connected with (every or) many others discourages partial or piecemeal social change. There are two faces of this tendency firstly it may be a conservative view to discourage *any* change, to avoid a wholesale social instability if not disorder. Secondly, it may justify revolution or wholesale social change to amend or improve some local/partial social malady. Incidentally Mill's actual social and political doctrines aim at making piecemeal social reforms and change instead of advocating holistic change.[42]

The objection on the ground of having an 'absolutist' view of trend is purely methodological. Popper agrees with Mill that there *are* identical trends found in different periods of history in different societies. But he disagrees with Mill on the point of this trend being any law, but only a singular historical statement. Laws and trends are different; the former does not depend on any initial conditions whereas the latter does.[43] Taking trends as laws is "absolutist" move in Popper's terminology. Indeed Mill gives this impression because his use of 'empirical laws' is quite indeterminate ranging from genuine low-level laws to descriptions of observed uniformities. It is unfortunate that Mill's phrases are not very precise, but he is aware that trends or statements of observed facts need initial conditions to have predictive value. But as I mentioned earlier, the existence of a trend or a tradition, in his view is a guidance to understand future social phenomena, although it is short of the status of law, which only can have predictive power. But like all classical economists he believes in some general laws of economics that are fundamental like natural laws. Still he advocates reversal of many social trends for egalitarian purposes. This is possible only if we have knowledge of the trends because then we have the power to bring forth social change, i.e., reverse an existing trend.

---

[41] K. Popper, *Poverty of Historicism.*

[42] Alan Ryan rightly pointed out this aspect of Mill's political principles.

[43] This is a major point where the institutional economists breakaway from the classical economists who hold that economic laws are applicable to *all* economic situations, while most institutional economists stress the historic, systemic and evolutionary nature of economic processes.

The last but not at all the least important is the criticism based on the assumption that Mill believes naively in 'progress as an unconditional trend'. We have seen before that Popper accuses Mill of 'psychologism' and particularly the belief in progressiveness of human nature. But as it is mentioned before, progress or progressiveness in this context is never meant in the sense of improvement or development, but only meant in the sense of cumulative effect of knowledge and achievements. In the essay *Civilization*, Mill made a distinction between two senses of the term[44]; in the sense of better and in the other it means to have more wealth and power. He is aware of a nation's being "civilized" in a wider or in a narrow sense.[45] It means wealth and power in the narrow sense and moral and intellectual achievement in the wider sense. That the present age is 'civilized' in the narrow sense is clear, but Mill says, 'we do not regard the age as either equally advanced *or equally progressive*. In some it appears to us stationary, in some even retrograde'[46].

Popper seemed to overlook these distinctions Mill made between different meanings of "civilization" and "progress" and only emphasized Mill's ambitious picture of rationality in science and extension of that to social phenomena. Surely inevitability of social progress or prediction of social events from trends is not a viable programme as Mill projected. But ascribing "historicism" or "historicist prophecy" to Mill's social philosophy would be incorrect and also not in tune with his ideal of individual freedom. But that ideal as shown below is in direct opposition to his project of social physics.

## *Social law and individual freedom*

The liberty of the individual, in Mill's view is threatened not so much by open attacks as by the quite erosion of liberty brought about largely unintentionally by moralists and political reformers. In fact Mill's *On Liberty* is a response to a danger - danger already envisaged by Alexis de Tocqueville; (in his *Democracy in America*) namely the danger not of undemocratic *laws* or other overt and political measures of an illiberal tendency, but of a more subtle kind of tyranny: the tyranny of a popular ethos over personal beliefs. 'The "tyranny of the majority" not over the body but over the mind'.

The threat of social tyranny is very real in societies like class-ridden society of England in Mill's time or India's caste ridden system. 'A social tyranny' as Mill rightly projects is 'more formidable than many kinds of political oppression

---

[44] I owe this point to A. Ryan.

[45] J. S. Mill, *Dissertations and Discussions*, Vol. I. p. 171-72.

[46] Ibid., p. 161.

since though not usually upheld by such extreme penalties, it leaves fewer means of escape, penetrating much more deeply into the details of life and enslaving the soul itself'.

The great difference between political freedom and social freedom has been aptly described by R. B. McCallum in this way[47]: a reasonably affluent young man of modern time will find himself back in the nineteenth century in an easy life with light burden of taxation, licenses, without security and other travel regulations which we face now. On the other hand in many matters affecting his private and intellectual life, views on science, religion and sex 'he would find himself bound by a constricting orthodoxy' and would have to guard his words and actions which he might find intolerable. The main thrust of *On Liberty* is to show the way the powerful Victorian ethos operate in a quite and comparatively painless way achieving its effect without the unpleasant process of fining, imprisoning or killing anybody. It even allowed dissenters to 'exercise their reason' at least in private provided they disguise their heterodox opinions. It is a 'convenient plan for having place in the intellectual world'. But the price paid for such intellectual pacification is the sacrifice of the 'entire moral courage of the human mind'. This "yoke of opinion" had become so heavy as to thwart independent and energetic action. Mill predicted that when democratic majorities have come to feel their power they would be tempted to use it to excess, so that civil liberty will be no less invaded by government action than social liberty had been invaded by social opinion.

Whatever is stated so far about Mill's doctrine of individual freedom is well known. One important point however needs to be mentioned here that his libertarianism does not invoke any abstract right to liberty. Individual freedom is not a natural right, "as a thing independent of *utility*". The central tenet of utilitarianism is that everything should be judged according to whether or not it tends to promote the greatest happiness of the greatest number. But there is another principle (briefly suggested before as anti-paternalistic), namely the principle of non-interference according to which society must not interfere with an adult person merely for her own good even if by doing that she would be happier and no one else less happy. Mill's "very simple principle" popularly known as the "harm principle" does not appear to safeguard the liberty of an individual when her actions significantly affect other people, surely what we say and write may affect others adversely.

The problems inherent in the harm principle generated much heat and literature over the years and it seems that the principle appears either to label all *actions* as harmful or leads to illiberalism. Without going into that controversial issue, we should refer to another principle, which is not given much attention by

[47] R. B. McCallum, ed., *On Liberty*, Introduction, p. XIV, cited by J.W.N. Watkins, 'John Stuart Mill and the Liberty of the Individual', *Political Ideas*, edited by D. Thompson.

liberals and others. I shall call this principle the principle of fallibility or rationality. This principle is a very general principle in the sense that liberal principles in general can follow from it. Mill's insight should be duly recognized for it provides a general ground for individual freedom independent of the harm principle. He argues that freedom of thought and freedom of expression are quite inseparable. But the principle of non-interference and harm principle *only* ensures freedom of thought. But if thought and expression are inseparable then an illiberal might argue that at least *some* expressions should be coerced or banned because these might cause some social harm. And once some expressions are branded as harmful, many others may also be banned for potential harmfulness by illiberal utilitarians. Examples abound like banning some book in advance in anticipation of some social unrest.

So the principles of non-interference and the very "simple principle" alone cannot safeguard individual freedom. Liberty of thought and expression therefore need another principle to be justified on rational ground. And that principle is based on fallibility of human reason. Let us look at some possible arguments[48] based on that principle. One: only trivial truths are certainly known. All other truths are contingent, that is dependent on circumstantial evidences therefore for progress of knowledge or amount of truth we need to have more and more discussion. Except logic and mathematics, we cannot know of an opinion (a belief based on empirical data) that it is true simply by looking at it. Moreover we cannot rely on feeling of *certainty* about any idea, because an opinion may be false even if we feel *certain* about its truth.

What entitles us to hold an opinion with some confidence can only be the knowledge that it has so far withstood severe criticism. Thus even in the case where a received opinion is in fact true, and a heretical opinion opposed to it is false; it would still be injurious to truth to suppress the false opinion. To do so would deprive us of an opportunity to test the true opinion; and the less a true opinion has been tested, the less reason we have for assuming it to be true and the less we appreciate its internal resources, its strength and significance. When any opinion (even if it is true) is protected against criticism it subverts into dogma and intellectual progress halts or at least slow down. Moreover it is irrational to cling to an opinion if one is ignorant or has been kept uninformed or ignorant of counter-opinion or rival theories. "He who knows only his own side of the case knows little of that, – he has no ground for preferring either opinion" (*On Liberty*). Surely any victory would be vacuous and meaningless if there is no enemy to fight in the field. Two: Of course it may be the case that the heretical opinion is true and the received opinion false - or, what is more possible both

---

[48] The rudiments of these arguments are already present in Mill's *On Liberty*, where he is discussing freedom of thought and expression'. I have expanded it into full-fledged arguments.

opinions may be *partly* true and *partly* false. A public censor would be unable to know in advance whether the received opinion is *wholly* true, since it is only *after* opinions have been allowed to collide that we can begin to assess their respective strength and weaknesses. As Mill said 'There is the greatest difference ... for the purpose of not permitting its refutation'.[49]

Three: As Watkins pointed out, *On Liberty* was published in the same year as Darwin' *The Origin of Species*, and one might say that Mill regarded the growth of knowledge as a sort of evolutionary process in which true opinions tend to grow stronger and to multiply - provided the struggle for survival is fierce. It was with his idea of intellectual progress towards truth that Mill backed up his famous slogan that the peculiar evil of silencing the opinion would be justified in silencing mankind.

Apart from his arguments for freedom of speech and action, Mill's commendation for individuality itself can stand against his principles for social mechanics. Mill's endorsement of one who is self-made is surely contrary to his vindication of social physics and more in tune with the principles of liberty espoused in *On Liberty*. He upholds someone who has formed her own opinions and worked out her plan of life rather than one whose opinions are not original and life-style according to customary norms. If customs and traditions are kind of social laws as we find in *Systems of Logic*, then Mill's applause of eccentricity (even if it is for its own sake than in pursuit of some distinctive individual aim) - seems to be incongruous. Indeed Mill approved of eccentricity indulged in not only for its own sake but also rather for its custom-disturbing effect. The only reason for endorsing eccentricity rather than blind perusal of customs is that 'the tyranny of opinion makes eccentricity a reproach'. In a society in which individuals lead their own distinctive life according to their own peculiar tastes and talents - freedom and reason seem to dominate more than trends and traditions.

But individuality to Mill is not as fundamental as free speech. If someone's action hurts other people only then that action be restrained, but if someone lives without hurting anyone and at the same time not very distinctively, or on the contrary in an innocuous and customary manner - we are not entitled to use moral pressure to force him out of it. If one emphasizes that people should not be a conformist and imitative as Mill did, it would be ignoring the immense service custom with its readymade norms does to individual behaviour. Mill first attempted a science of society that follows the laws of tradition based on universal laws of human nature. Then the principle of fallibility of human reason ensures freedom of speech and action. Finally the concept of individuality acts as an antidote to the effects of customs and social mores to our thought and action. It is true that all our customary modes of behaviour are not reasonable and most need improvement.

---

[49] J. S. Mill, *On Liberty*, Chapt. II.

However one should also remember that reform of one practice or institution requires reliance on a great variety of practices and institutions. The stability of the latter institutions is likely to be indispensable for the rational reform of the institution in question. One cannot just defy all customs; by defying one custom people may turn to another custom. It is replacement and not mere rejection of *all* customs whatsoever. Mill seems to ascribe almost nothing to the individual beyond her sovereignty over her private thought and feelings. Of course he does make a distinction between conduct that does and conduct that does not affect the interests of others prejudicially. But that again is problematic for one's action can affect other's interests prejudicially, but he cannot be said to have caused them definite damages and vice versa.

Moreover Mill made another condition for control of freedom of speech and publication; namely there are occasions when freedom of speech may legitimately be curbed. He would say it is one thing to publish an article in a news paper which argues for example that hoarders of food grain make people starve by artificially creating a price rise and another thing to make a speech with the same content when angry mob assembled in front of a grain-hoarder's house. Such restrictions put on the exercise of freedom of speech and action makes such freedom innocuous and meaningless.

John Watkins[50] made a distinction of different states of heteronomy of mental and physical conditions where autonomy of the individual is absent in various degree and kind. Some time it might also be self-inflicted. He referred to the case of Katherine Mansfield's *The Daughters of the Colonel* where even after the death of their father the daughters remained in his thrall. Both Mill and the critical rationalist may agree with Kant and his enlightenment ideal in this connection. Especially when the latter declares that only a lazy mind does not use its own judgment and reason and depend on others for the course of action to follow. Critical rationalists similarly thinks that an action that shows high degree of freedom and autonomy also shows high degree of reason and thought. Watkins finds similarity between his idea of autonomy and that of Mill. He says: ' Like Kant who declared that reason can accord sincere respect only to that which has been able to sustain the test of free and open examination (*Critique of Pure Reason, A* xi), Mill was in many ways a good critical rationalist. His *Liberty* (1859) broke with the classical empiricist tradition in which his *Logic* (1843) was embedded, especially in repudiating the no-invention-of-ideas thesis: 'nothing was ever yet done which some one was not the first to do, and ...all good things are the fruits of originality (*Liberty*, p. 268)'.[51] What is needed for intellectual (and moral) progress is bold new ideas plus fearless criticism. An opinion should be

[50] John Watkins, *Freedom after Darwin*, chapter 8.
[51] Ibid.

accepted, though only tentatively, when 'with every opportunity for contesting it, it has not been refuted'.

Mill's idea of individuality was like Kant's idea of enlightenment in calling for movement away from certain negative states. Its attainment calls for a strong will, mental vigour, moral courage and energy (p.272), which sounds like a position well up on Watkins' disonomy scale. Mill said that the lives of people who lack individuality are 'pinched and hidebound' and 'cramped and dwarfed' (p. 265), which sounds like our anonymous heteronomy. He also wrote: 'In our times, from the highest class of society down to the lowest, every one lives as under the eye of a hostile and dreaded censorship' (264), which sounds like outward conformism. But outward conformism may slide over into inward conformism, and Mill also declared that society 'practises a social tyranny more formidable than many kinds of political oppression, since, though not usually upheld by such extreme penalties, it leaves fewer means of escape, penetrating much more deeply into the details of life, and *enslaving the soul itself*' (p.220) My italics). That sounds like internal heteronomy.'[52]

It may be mentioned in this connection that Mill also offers an argument to defend freedom of speech which is based on fallibility of our beliefs. The argument runs like this: we should not suppress any opponent view (or action) because all our beliefs are fallible. Therefore the opponent view-however absurd it may seem – might very well be true and a long-held belief false. All opinions should therefore be expressed. The similar rapproachment between Mill's notion of liberty of thought and speech and critical rationalism is apparent from his fallibility of reason argument expounded in *Liberty.*

Going back to the issue of making provocative speech or action that may flare up dangerous repercussions leading to violence and other harmful consequences, should there be a pre-emptive measure to stop that? One may argue nevertheless that a true champion of freedom of speech would not put any restriction, nor incur any punishment for preaching any so-called provocative speech. Because then any speech or act might have been pre-emptively stopped in the name of prevention of future harm. But we can never ascertain which speech/writing will lead to harm. Even *after* a disaster has actually happened, it is not an easy task to assign its *exact* cause. Moreover publication of a book or scientific paper that *might* lead to disastrous results then can also be justified to be prohibited. It is true that historically many events such as the content of Einstein, Bohr and Oppenheimer's scientific papers eventually led to dangerous repercussions; but that does not justify that such scientific discoveries should not have been published. Science presumes a free society. A free-society is one which does not deny the role of tradition but nevertheless allows free thought and its expressions, no matter what repercussions it might bring forth. It is rea-

---

[52] Ibid. p. 228.

sonable to presume that the pursuit of truth is a risky business because it may yield large benefits to society or may also lead to dangerous discoveries. And therefore should we control human inquiries so that they only yield benefits and no harm? But alas! No pursuit of truth can guarantee only *benefit* and no harm. So it would be a question of security vis a vis knowledge, which is the issue and it leads to a tussle between liberty and security. If we value security then freedom of all kinds has to be compromised. But if liberty as a principle is valued as such then the cost of it is very high because freedom does not ensure security of any kind. The risk of discovering unpleasant truth lurks behind any free pursuit of truth.

With the breakdown of determinism in contemporary physics we should have a fresh look at the question of human freedom and those who think that physical determinism is reconcilable with it. Freedom as we will soon see is a kind of emancipation by the use of reason, from dependence on external factors such as social, historical or even psychological laws.

# *CHAPTER II*

## FREEDOM AND EQUALITY

'All men are equal, and some men are more equal than others'. George Orwell

Freedom and equality are often regarded as two different types of social ideals that cannot be realized together. I shall try to show by analyzing these two philosophical concepts that as social ideals they are not as incompatible as they look. The question of human freedom has been approached in two opposite ways by the classical empiricists and the classical rationalists. As for the critical rationalists Popper gave more importance to critical rationalism as an epistemological theory. Watkins[1] for the first time offered an elaborate theory of human freedom from the critical rationalist point of view which made a distinction among different states of freedom and the degree of overcoming stages of unfreedom. We shall come back to that later in the discussion.

We have already discussed Mill's argument for 'absolute freedom of opinion and sentiment on all subjects, practical or speculative, scientific, moral or theological'. But it is clear that there are occasions on which liberty of publication, expression and association could be overridden where they led to demonstrable and unequivocal 'harm' to others. In other words, liberty of expression and association is a prima facie right of the individual but that right can be circumscribed under certain well-defined conditions.

We have seen that the principle of liberty is weakened by its 'harm' principle. So it needs some other principles as a support. Unqualified freedom cannot be pursued in any civil society. Two other principles, namely the principle of utility and the principle of fallibility of reason are at par with freedom of expression and association, but the principle of utility is not necessarily related to either freedom or fallibility of reason. We can, for example, think of a closed society where individual right of free expression is not recognized nor is it felt that some opinions (such as the ones held by the authority) may be false - but social goods are handed down according to the need of the people. Of course one may question whether tailored utilities without freedom can maximize happiness. But well-oiled social machinery might condition people also to complete complacence with a particular kind of state of things. Anyway it may be contestable whether utilitarianism as such is a self-sufficient principle to be valued or it needs other principles to support - but one thing at least seems to be clear that principle of utilitarianism does not necessarily need a free society. In other words

[1] John Watkins, *Human Freedom after Darwin.*

freedom and utility seem to be two independent concepts. But before arguing for this claim let us look at some historically important interpretations of problems related to freedom, equality, and justice as social goods.

## *'Just Society and Freedom'*

In his seminal work[2] John Rawls proposes that in a "nearly just" society principles of individual liberty may always be taken to override principles of equality. His key notion is that one may deduce one's political and moral obligation from the notion of rational self-interest. The idea of placing people in the 'original position' is a metaphor for Rawls' concept of human rationality. A rational man would opt for a form of society where he would be better off and better protected than in other societies. Rawls thinks that he would primarily seek the good of protecting his liberty. Interference is not justifiable by an appeal to countervailing demands to ameliorate poverty or to enforce a more egalitarian distribution of goods.

But *equality* also is important for Rawls who labels it as the *Principle of Difference; a rational man* (though with a "veil of ignorance" about his fate in future) would not take any risk with his own future prospects. He would opt for a form of society in which even if he occupies the lowest rank, would be better off than alternative societies. On an almost similar voice to Mill, he pronounces 'each person is to have an equal right to the most extensive total system of equal basic liberties compatible with a similar liberty for all'.'[3] So freedom gets the first priority, but equality is also important; the 'principle of difference' as he calls it, is less radical than many new version of egalitarianism. Equality is first the normative baseline. All inequalities therefore need justification on rational grounds, although he does not consider 'classless society a possibility' for a rational agent to opt for.

Rawls's Principle of Difference (equality) contains two components: The first one assumes for the rational agent not to take any unnecessary risk for the future. They act according to a 'maximizing principle which ensures that the 'least advantaged class' in a society ought to be protected. This amounts to saying that in his account a class-ridden society is only permissible or justified if it is both a precondition of and has the effect of improving the lot of the poorest members. It follows therefore that Rawls would prefer a stratified society in which the poor are reasonably well provided for to a more egalitarian (or free) society that allows or entails poverty for its worst-off class. This part of his principle of difference makes again the existence of a better-off class essential for raising the level of the worst-off class. This assumption suffers from a non-

[2] J. Rawls, A *Theory of Justice.*

[3] Ibid., p. 60.

egalitarian stance contrary to the claim of the principle of equality. Moreover 'the trickle down effect' argument can also be justified on this ground. The argument rationalizes more money, more opportunities for rich/privileged people so those poorer ones also enjoy the trickled down benefits from the better-off class in the long run. The argument tries to justify inequality on the basis of justice and begs the question. Moreover it overlooks the essential paradox that instead of social equality it needs the rich/privileged to be more so thus defeating the very purpose (equality) it stands for.

Karl Popper regards the principle of difference as an attempt for a greater equality among people and their income to be not very far from the socialist ideal. The latter in his view involves lot of problems. First of all socialism is not a very efficient system in economic affairs, and shall multiply bureaucratic complications and more importantly it is 'intrinsically illiberal'. He thinks the attempt to reconcile equality and liberty as proposed by Rawls in his principle of difference is not a functional concept. Moreover the practical consequences of a similar theoretical attitude shall bring a general decline and crisis of democracy.[4] If the existence of distributive inequality can be justified on the basis of protecting the worse-off in the long run, then the principle of difference becomes trivial because it can be satisfied in many kind of society - (a) societies where there are no inequalities, (b) societies where there are great inequalities and also (c) societies with inequalities of different shades.[5] In brief Rawls's account allows that inequality in principle can always be justified.

The second component of the Principle of Difference however has priority over the first. According to this principle, social and economic inequalities are to be arranged so that they are 'attached to offices and positions, open to all under conditions of fair equality of opportunity.[6] The second principle demands justice as fairness to all. It does not simply allow (like the first one) that inequality in society is justified only if the least advantaged class benefit from it. It positively excludes, denying or restricting opportunities to other members of the society in order to improve the lot of the underprivileged. Thus this principle would presumably prima facie block reverse discrimination type of social inequality. It also implies that affirmative actions (such as quota or reservation of seats or posts for special classes of society) or reverse discrimination can *only* be justified if there is an existing injustice or imbalance in the distribution of social opportunities which needs correction. Fairness or justice therefore requires rehabilitation of underprivileged by granting them extra opportunities that might deprive other sections of the society of similar privileges. Without the presumption of prior 'unfair' distribution of opportunities - inequality of opportunities can never be

---

[4] 'Better be Liberals than the Greens': Karl Popper's interview by Sebastiano Maffetoni in Italian, *Panorama* 1989.

[5] Ted Honderich , *Three Essays on Political Violence.*

[6] J. Rawls , *op. cit.*, p. 83.

justified. According to this principle reservation, quota or special help in educational institutions and offices can be justified only for the first generation underprivileged, but once a special assistance is enjoyed by a member of some community or social class - such affirmative action to be continued *after* that generation will amount to unfair distribution of opportunity.

We see thus by combining these two components of the principle of equality Rawls ascribes first priority to liberty.[7] Equality of opportunity and improving the lot of the worst class get only a second berth. He has been adequately read and criticized from various standpoints. Without going into those squabbles one point may be emphasized that his idea of equality as 'fair opportunity' is a pale shadow of his concept of liberty. The latter is positive and independent of the idea of equality which however seems to be a secondary concept dependent on the principle of basic liberty for all. Equality as such is not justice - it is 'fair opportunity' or particularly remedying poverty. Surely there is more to equality than just improving the lot of the worst-off. Poverty is not the only menace of inequality as the latter does not necessarily include poverty. Moreover, 'poverty' again means different things in different societies. U.N.O. might have a specific definition for it, but still different countries have different economic model depending on the variant kind of initial conditions they involve. Although the concept of equality may involve attempt to eradicate unfair distribution of resources, the basic idea of it includes recognition of the undesirability of a class-ridden society. Rawls's notion of rational man rejects this intuitive aspect of equality and includes on the other hand a notion of rationality which requires members of a society to prefer a state of maximum material benefit for themselves and it is unreasonable for them to prefer less prosperous life in the interests of benefit for less privileged class. Indignities of a hierarchical society are missed by his overbearing notion of freedom. For Rawls inequality is justifiable so long as it provides for the poor or the lower order. This model of equality thus makes rationality a concept independent of any moral question although the principle of difference tries to amend that to some extent, but not with much success. Justice-based concept of inequality can have many adverse social consequences. Thus Rawlsian type of egalitarianism is innocuous and subject to misinterpretation and misuse.

---

7 A. K. Sen argues that it is a gross oversimplification to assume that individual utility functions are more or less the same. Everyone *cannot* derive same amount of satisfaction from various things due to many reasons (handicap or sheer luck), *On Economic Inequality*. He also emphasizes the importance of freedom in determining standard of living. Freedom means having a capability set that offers plenty of opportunity for choice. He further refines the concept by distinguishing 'agency freedom' from 'well being freedom'. Sen, 'Well-being, Agency and Freedom', *Journal of Philosophy*, 82, 1985.

Justice-based Equality Egalitarianism may have, contrary to the above view, a moral version. It asserts[8] that equality is preferable in its own right and more importantly existing inequalities stand in the way of human self-realization and autonomy. Equality in this sense is a positive concept; it brings about "human flourishing". It is a positive concept (somewhat similar to the positive concept of freedom) in the sense that it entails not only the absence of constraints upon what one desires but further, the incorporation of a moral ideal. The ideal is the vision of a just society in which people are encouraged to pursue intellectual aesthetic and physical activities for their own sake within a co-operative, humane and unselfish framework . Contrary to Rawlsian vision this view does not allow inequality to be justified, does not even require treating equals and unequals differently. Nielsen's version does neither envisage a faceless society where sameness prevails over variety and individual difference nor does it recommend uniform life for all the members shorn of distinctive style and eccentricity (which Mill dreaded so much). It recommends on the contrary that removal of inequalities may gradually enable individuals to pursue their own interests and life plans. The two major principles of such egalitarianism are as follows:[9]

(1) "Each person is to have an equal right to the most extensive total system of equal basic liberties and opportunities (including opportunities for meaningful work, for self-determination and political participation) compatible with a similar treatment of all.

(2) After provisions are made for common social values, for capital overhead to preserve the society's productive capacity and allowances are made for differing unmanipulated needs and preferences, the income and wealth (the common stock of means) is to be so divided that each person will have a right to an equal share. The necessary burdens requisite to enhance well-being are also to be equally shared, subject of course, to limitation by differing abilities and differing situations (attributable 'to) natural environment not class positions".

It is clear that (1) and (2) together look to be quite comprehensive and suggestive of ways to implement them. But the problem is that egalitarianism presupposes a kind of condition or milieu which no existing society can provide for. Nielsen suggests that in such society individuals would live with reasonable affluence, have a secure life and they would not be possessive of resources nor greedy selfish human beings. The aim of their economic institutions would not be profit maximization but 'the satisfaction of human needs of everyone'. Industrial and political policies of such a (democratic) system would be cooperation rather than competition.

How does this society implement the principle of just distribution without infringing the individual right to a free and distinctive life from others? In a

---

[8] Kai Nielsen holds such view. See his 'Impediments to Radical Egalitarianism', in *American Philosophical Quarterly*, vol. 18, no. 2, April 1981.

[9] K. Nielsen, Ibid. p. 121.

totalitarian system freedom is not respected so just distribution would be uniformly same for all somewhat like a military mess or a refugee camp. The eternal question (immortalized in Dickens' *Oliver Twist*) 'Sir can I have some more?' does not arise there at all. But in a democratic society how can a just distribution principle work? It requires equal respect for everyone's special need and special talent. It is recommended first that *all basic civil liberties* would be guaranteed and protected without reference to class position. "Social status" system should be abolished or minimized and respect and honour would be conferred to individuals and not attached to any social roles such as principals, managers, ministers or doctors. The problem here is that individuals achieve respectability by assuming some social role, so it is difficult to dissociate the social and the individual in one particular person. The spirit of such idea however is understandable - to elevate the individual beyond institutions. Another difficulty is that all societies have some history. And social status, honour, respect etc. should be understood in that historical perspective. It is true that many such *traditions* are illiberal and undermine individual person's respectability. Royal/rulers family or Brahmins for example were respected at one time (not always irrationally) but with changes in society that kind of respectability looks unreasonable. In an open society individuals should be valued for their individual abilities and not for their inherited status or caste. A society is usually regarded open according to the degree of value it attaches to persons for their individual abilities rather than for their social roles. Another requirement for a perfect egalitarian system is that there should be a 'common stock' from which certain basic commitments (to satisfy some basic needs of people) could be fulfilled. Education, healthcare, pensions, etc. are such basic areas. After satisfying these needs resulting in moderate abundance, Nielsen proposed that the residual resource should be utilized for people's "unmanipulated needs and preferences".

The idea here is that egalitarian society should not presume that individuals differ in their ability, need and talent, these differences should not be undermined, nor sustained by a mass regimentation machinery because in any egalitarian society people will at least differ in their individual aspirations and talents. Such needs would not be results of any social determination but flow from personal preferences linked up with people's distinctive aims and life plans. Given that individuals differ in their interests and capacities, each would receive from the common stock only what she needed. As individuals differ in their interests and capacities, and individual needs are not equal, it follows that some would take more than others would and some less. The criteria for (just or fair) distribution therefore would be not (only) according to need, but according to capacity, i.e., according to ability to benefit.[10] It is true that the present social arrangement

[10] The same units of utility-value may not bring the same amount of benefit to different individuals. See A. K. Sen, 'Rawls vs. Bentham: An Axiomatic Examination of the Pure Distributive Problem', first published in *Theory and Decisions* IV, 1974; also in N.

even in the most open and democratic societies is such that more often than not the "ability to benefit" from say, education, healthcare, or other opportunities are class-biased. For example the actual conditions are such that women, non-whites and other underprivileged sections cannot make best use of many opportunities due to complex reasons (depending on the particular social conditions) such as social taboo, lack of information or simply for biological and other factors. But nevertheless we can identify some talents, knacks or capacities "naturally" or "genetically" given to individuals. The common stock should encourage and sponsor these abilities and talents. Talented individuals require more attention as well as should carry more responsibility for the social good.

It seems that this kind of egalitarian distributive system with its complex structure to adequately assess individual needs and aspirations requires highly centralized bureaucratic machinery. But surprisingly Neilsen does not think so. In his view a 'co-operative' system of distribution with many smaller democratic institutions within it can make it successfully operative. He admits however that the complex social conflicts would not completely disappear but be minimized or become redundant. An analogous case can be minimization and in some cases redundancy of "black-marketing", "smuggling" or "bribing" by free economic measures such as easy import policy or discontinuation of strict license/permit policy for trade. To value such an egalitarian programme is difficult for even if it is morally correct, implementation of it in just measure is difficult. Because there are so many complex factors involved those chances of misuse and mistake, injustice and oversight are very high. But surely a philosophical principle should be valued not only for its practical value but also for its logic and internal consistency. The most striking feature of Neilsen's model is its moral assumption. Rawls's rational man and Neilsen's man with moral will are comparable as to the value of their ideal, Neilsen's society with good will of all is also not an actual society, but one envisaged to fulfill basic human needs as well as one in which individual goals can be pursued to the maximum without hindering similar goals of others. The moral strength of such a (distributively) just society is undoubtedly great.

But it has some internal problems in addition to its being not an actual society but a hypothetical one that satisfies the requirement of being a society of individuals with a fair amount of "good will" analogous to Rawls's requirement of "rationality". The other problems are first, if we allow the 'common stock' to be used primarily for basic needs - civil liberties may be suspended temporarily in a transitional period to meet them. But basic needs being arbitrary and ambiguous - the suspension of free use may continue indefinitely like any emergency situation such as war, flood or civil unrest. There is also a possibility to misuse the so-called fulfillment of basic needs permanently in an unfree manner.

---

Daniels (ed.), *Reading Rawls*. See also 'Equality: On Sen's Weak Equity Axiom' James Griffin, *Mind* (1981), vol. XC, pp. 280-286.

Any imposition of emergency laws (such as "the long emergency period" in India imposed by Indira Gandhi government in the 'seventies) can exemplify that. It might also entrench indoctrination or other forms of social control, injustice and coercion in the name of social good. Neilsen may argue that justice, as equality, needs to be realized by 'moral argumentation' in the public sphere. He is aware that such argumentation would be resisted but holds nevertheless that such resistance would be preferable to violent change like a revolution or civil unrest. Another problem this picture of decentralized power of social institutions faces is its utopian character. The hard-left champions of equality such as some Marxists would regard it as impractical, unrealistic ideology that does not recognize the actual factors of social processes such as social, psychological and other trends and put too much importance to the presumption of sense or "goodwill" in human beings. It is true that this view does not even count the self-interested nature of people. But then a philosophical prescription is different from a sociological account. He has only proposed a possible picture of an egalitarian society. Another internal problem of the picture is its emphasis on 'moderate abundance', affluence etc. The picture might have been workable if we know for sure what abundance, affluence and its degrees mean. Of poverty, however, United Nations is trying to standardize its implication. But that standard cannot clearly indicate what would be abundance.

Moreover an egalitarian picture of society needs to accept that inequalities exist and that they are obstacles to the concept of a just society. Identifying different kind of inequalities and suggesting ways of remedying them is a burden on them. A full-fledged egalitarian programme with sociological explanations is not necessarily required. But accepting inequalities of different kinds, an egalitarian view needs arguments to show why these are wrong. That cannot be achieved just by visualizing what an egalitarian society would look like. Many feminist philosophers for example are doing such job of identifying areas where gender-inequalities exist along with arguments why such inequalities should *not* exist. We shall come back to this issue later in connection with the interlink between freedom and equality.

To summarize the difficulties briefly: (a) an egalitarian should address the question why greater equality is desirable than inequality or lesser equality? What is the ground for assuming apriori that all inequalities are prima facie wrong? As I mentioned before we have a long history of various kinds of inequality some of which are based on good ground and some are not. (b) If the actual existence of inequalities is recognized, the process of correcting that disvalue by indoctrination or other methods involve lot of social control and coercion that squashes a great amount of individual freedom. So is it worth the tortuous effort? (c) Some may argue that inequality is inevitable and all forms of it are not necessarily harmful (d) without being a subscriber to Platonic view of hierarchical society - one can hold that some kind of inequality is essential for social stability and practical functioning of any system. It is economic for any system that works to have a hierarchy of not only fortune or pay scale but also of responsibility, talent and capacity. A subordinate employee for example enjoys less fortune or 'abundance' than one who is in charge. But that kind of inequality (say in pay and fringe benefits) does not necessarily generate resentment nor is an instance of unjust distribution because the 'superior' officer has correspondingly more demanding responsibility and risks than the subordinate person does. Unless roused by political or trade Union enthusiasts - people (unless there is a tremendous injustice involved or personal envy prevails) in general are quite adaptable to such social arrangements.

Considering that inequality is in some sense inevitable and not always unjustifiable, a question arises at this juncture: Is equality *per se* a social good or guiding principle? Is it so sacrosanct that individual freedom/liberty can be a secondary principle? If the answers are in the affirmative then one needs some arguments to show that. In this connection let us consider the view[11] that equality does not have any intrinsic merit. The argument runs as follows: egalitarianism assumes that there is something wrong with all existing distribution of goods and services. The interest is more on redistribution than distribution. The underlying assumption is unwarranted. Nozick suggests that it is more reasonable to see existing systems of distribution in a historical rather than a utopian perspective. The very conception of unequal distribution in this context is inaccurate for it assumes as if there is a central stock (and distributing power) from which resources are "doled out" to individuals according to some unspecified but unjust criterion: 'We are not in the position of children who have been given portions of pie by someone who now makes last minute adjustments to rectify careless cutting'. The quotation reveals that any kind of egalitarianism seems to imply at the best paternalism, i.e., a kind of protectionism that is directly a denial of individual freedom. In its worst form then egalitarianism may lead to dictatorship. Because as we have seen earlier, equal redistribution involves so many complex factors that it might need coercion and imposition on individuals free action for the collective social good. And if equality is the primary principle, freedom can take a second berth. Moreover no state has achieved this degree of control over the distributive process. Production and distribution in different societies have different historical background with different criteria of entitlement. And all entitlements are not prima facie unjust.

In this context Nozick[12] defines the 'appropriation of unheld things' as 'holdings', it may include property, personal and corporate and other less tangible social goods. Nozick argues that we can make a distinction between holdings acquired justly or not. There is nothing necessarily or prima facie illegitimate in certain individuals or groups acquiring or transferring holdings - the whole process can be done justly or unjustly. It will be an illegitimate holding if theft fraud

[11] Robert Nozick developed such a view in *Anarchy, State and Utopia.*

[12] Ibid., pp. 58-79.

or robbery acquires it. If I obey the rules (like property law, or other laws of the state) on the other hand - my entitlement would be legitimate although it makes me a little unequal to many others.

By this kind of argument thus, maxim of justice should *not* imply that: holdings ought to be equal, unless there is a good reason why it should not be so. It would rather be that people are entitled to keep their holding (if acquired legitimately) unless there is a good reason why they should not do so. The onus is thus on the egalitarians who regard that resources should be redistributed in accordance with the principle of equality.

Two other questions also become relevant to be addressed, one: Can we frame a formal principle of equality that can be embraced by all egalitarians in the same way as the formal principle of liberty is acceptable to all libertarians and *most* egalitarians? Two: Should one take account of actual concrete, historical context in which the principle of distribution is placed or else opt for an abstract theoretical principle of equality which Nozick regards as utopian and non-historical?

The second question can be taken first as Nozick's charge of utopianism to radical egalitarianism (of Nielsen's variety) seems to be the most serious objection presented against the principle of equality. It is true that any utopian society is subject to logical and empirical scrutiny, because an adequate blueprint of such a system is not easy. Utopian ideas such as ones of an egalitarian society need not give a detailed account of how it should be constructed. It may be viewed as the possible implementation of moral ideals. Gandhi visualized such a society as 'Ram Rajya' or the Divine (Ideal) State that fulfills the requirements of a just society. General theoretical moral concepts act as an ideal or model that still requires working out of the functional details. Thus the principle of freedom as absence of constraints and the principle of egalitarianism can be seen as embodying moral considerations which have to be taken into account in reaching a moral verdict. Taken separately, neither principle can dictate a verdict either political or moral. For example to put constraints or to abolish private educational institutions that serves *only* the rich and the elite, both principles need to be referred to for prima facie moral requirements.

These considerations also address our first question namely about formal framing of a general principle of equality (without reference to historical background). We saw above that we all refer to abstract principles for moral and political issues. Then Nozick's challenge that why accept the formal principle of equality at all? Or, why should we assume that prima facie, all people should be treated equally and given equal shares? This seems to be unwarranted. In other words, the burden is upon the egalitarians to show that the existing system of distribution is wrong and it calls for redistribution. It is true that in a sense it is arbitrary to say that given the social-historical tradition of a society 'whatever is *held* is held justifiably'. Since "being held" and "being held justifiably'" do not have much difference in meaning since given the socio-historical background all existing holdings can be shown to be justified barring a few exceptions such as fraud, stealing etc. But whatever the law of the society approves is not necessarily of moral worth. So the egalitarians are right in a sense that all (existing) distributive system requires that we invoke some moral maxims to justify them. And the principle of equality is a general principle from which such moral maxims should follow. Of course we cannot invoke a principle with monolithic maxims. The latter should take into account the variety of social arrangements, customs and criteria of entitlement.

But if we consider all entitlement to be justifiable - then the notion 'justifiable' soon become meaningless and the theory gets very close to cultural relativism. Of course there are variety of social arrangements of distribution but that does not mean all of these customs are equally justifiable. We can surely make a rational adjudication among different arrangements of holdings. For example distribution can be by inheritance in one system, according to hard labour in another or else according to alphabetical order of first names of the individuals in still another society. Surely these are not equally justifiable by any moral standard. In short these considerations show that if we allow that holdings or entitlement can be justified according to the norms of the tradition in which they are held, then we are close to cultural relativism and the rule becomes arbitrary.

The principle of equality on the contrary invokes an abstract rule that appeals to our intuition of natural justice, namely, the intuition that irrespective of cast, creed, birth, gender and age everybody should be treated equally contrasted with Nozick's or any other similar view that the notion of "just" holding is dependent on historical accidents - the principle of equality at least refers to a maxim that follows from a general principle. And the maxim also explains many existing democratic standards. For instance it is generally acceptable that high-income group should be taxed at a higher rate than the next high-income group and so on, and the lowest income group (the actual ceiling and the details only depend on the particular society) should be exempted from paying any income tax. Women and many other sections of the society again should get some special exemptions.

Similarly in educational institutions it appeals to any open mind that other things being equal all individuals should be treated by the same standard. A relativist who does not find any intrinsic value in the principle of equality may of course argue further that the norms the egalitarian thinks are natural or reasonable, are also contingent upon social conditions. For example she might say why "merit" or "poverty" are decisive norms, why not say "personal beauty" or "gentleness" of character can justify special treatment - other things being equal? The implicit suggestion behind this tirade is that these qualities are valued by society for long and there are no other rational grounds than that. But that leads to a further question, why societies in general uphold the values of 'merit' and 'poverty' more than 'personal beauty' and 'gentleness' in egalitarian decision? The inevitable answer would be that equality being the intrinsic norm of fairness - 'merit' and 'poverty' are justifiable grounds for any contingent deviation.

Another possible argument usually found refers to actual features of human life. It goes this way: inequality is an inevitable part of human society, it is evident from the standpoint of biology, human psychology and also supported by anthropology. In other words biological difference and psychological difference from one individual to another is such that it is almost senseless to say of two individuals that they are equal. Anthropological researches also show that people belonging to different societies not only differ in their physiological traits but also in their other traits, so it is difficult to find an uniform standard to judge them. Every human society is different from any other also in their social arrangements; therefore the concept of equality is meaningless. Every society operates through their standard of justice and equality.

The findings of socio-biological and anthropological research undoubtedly opened the floodgate of much new information so far unknown about different kind of people and their social arrangements. Without going into the details of these findings and accepting the basic truth of this kind of research, we can still say that all these findings can only show that one society and its people are different from others, but does not as such show that the different types of arrangements *cannot* be compared as to their moral worth. And an egalitarian does not need the premise that inequality prevails over societies. Because it only states that 'given the differences of race, caste, creed birth etc. everybody should be treated equally as to their social opportunities with exceptions only if there are some overriding reasons (such as merit or poverty depending on the nature of the case).'

But there is another formidable objection to egalitarianism in general which is based on some contingent factors. The objection amounts to saying that (a) inequalities are found to be contributing to social stability, (b) most resentment of people against inequality or injustice are low-range grievances aiming at the "reference group" they belong and not against *all* inequalities or against a group far away in the hierarchy.[13] This kind of relative deprivation 'comparable only to a narrow class leads to merely a minor state of hardship but the hierarchy remains unaffected. For example a daily labourer may feel injustice if his brethren in similar profession gets higher wages - but does never resent the high status and the enormously greater benefits enjoyed by say, the managing director. Runciman argues that society stays in this way stable. Social change can take

[13] W.G. Runciman has shown with the help of well-documented arguments that people's attitudes to objective social inequalities seldom correlates with the facts of their own position. He claims that people's resentment on equality usually ranges over their narrow 'reference group' and not over a much wider range, thus it helps to maintain social hierarchy. See his *Relative Deprivation and Social Justice*.

place only when questions are raised about injustice in a high scale across the hierarchy and feasible ways are suggested for such drastic change.

One more thing about egalitarianism and its enemies that needs to be explained before we pass on to the relation of the concept of equality to that of freedom which is suggested at the beginning of our discussion. But first about the possible difficulty faced by egalitarianism. The difficulty is whether egalitarianism endorses entitlement or holding. The fear that private property cannot be held appeals to the natural human instinct for privacy and ownership feeling. Neilsen[14] clearly ameliorates this fear by distinguishing between holdings like farms and homes built with labour and care that foster a sense of belonging with ... 'ownership and control of the major industries that is crucial'. It is evident therefore that it is not necessary for egalitarians to take away homes or other belongings from people. The only rider it requires is that entitlement can be justified only if it can be shown that the holding of the particular property is consistent with the removal of the more formidable forms of social inequality.

## *Freedom and Equality*

As mentioned in the beginning of this discussion freedom and equality as usually thought, are not two inconsistent notions. In fact the major reason for abolishing inequalities is that they operate as a constraint and hindrance upon people's individual will and aspirations. Inequality is disabling because it is an impediment to individual freedom. To make it more clear let us first turn to both the sense of 'freedom', namely positive and negative freedom. I shall like to argue that in both the senses freedom is a requirement on which equality depends for its justification.

Briefly speaking negative freedom is freedom from constraints and positive freedom implies that 'one is free to the extent of effectively determining one's own life', and obviously this does not only mean freedom from external constraints but an ability rationally to order one's preferences so that they do not run against each other and also not against one's basic purposes.

As we mentioned in the beginning, the word 'freedom' is usually used in various possible senses and not without an emotional overtone. The distinction made between negative and positive freedom covers the different aspects reasonably well. In a sense democracy allows people to do as they please without constraint, i.e., it encourages the development of personal autonomy. In this pursuit the exercise of freedom of choice is the highest value. Because given the rationality of an agent, the individual is the ultimate judge for what she wants to do for herself. The negative conception of freedom allows for the moral judge-

---

[14] K. Nielsen, *op. cit.*

ment that freedom seen as *absence of constraint* may be an end-in-itself. Of course freedom is only prima facie desirable, and there are occasions upon which it may justifiably be circumscribed. Mill's principle of liberty also contains this rider. Nevertheless in justifying constraint or interference with what people want to do, the burden is always upon the interfering power why freedom should be sacrificed for leading to a greater good. It is a highly controversial issue whether any democratic system, particularly democratic socialism would still stick to this notion of freedom if all the members of that society *want* to lead a lazy life with no enterprise for personal or social good, indulge in fraud, violence and many other undesirable activities or take resort to coercive measures for curbing such destructive attitudes. Such concept of freedom is obviously not sufficient because mere freedom from constraint can never ensure realization of one's goal since there are many complex requirements such as knowledge and information, facilities and right opportunities which play important role in one's realization of desired goal.

In this context the most important points are raised by Charles Taylor who argues that almost everything about the concept of negative freedom is not only naive but also wrong.[15] Taylor's main objection is that most versions of negative liberty overlooks the existence of a strong philosophical tradition embodied in theorists like Rousseau and Marx in which freedom at least involves "collective control of life" and is closely related with notions like "self-realization" or "rational self-mastery". Only to emphasize external constraints such as being 'totally unaware of his potential, if fulfilling has never even arisen as a question for him' or 'the fear of breaking with some norm which he has internalized but does not authentically reflect him'. So freedom implies not only freedom from external impediments but as mentioned before, an ability to rationally order one's desires so that they lead to the optimal realization of "ones basic purposes". Indeed the concept of "rational self-mastery" or "self-realization" smacks of idealism and is a suspect for encouraging paternalism, of different sorts. To know the 'real purpose of one's life' involves complex intangible factors no doubt, but it can still be said (as Taylor argues) that the meaning of freedom is necessarily related to "discrimination amongst motivations" or rational ordering of priorities.

The positive notion of freedom appeals to our intuition of a clear and sufficient meaning of the word "freedom" which surely implies more than freedom from external constraints. But one may also argue[16] that an analysis of the possi-

---

[15] Charles Taylor, 'What's wrong with negative liberty'?, A. Ryan edited, *The Idea of Freedom.*

[16] Keith Dixon has distinguished different categories of "constraints"(a) it may be intentional interference of other human agents, (b) it may be unintentional consequences from human action, (c) it may be our own desires such as (i) "irrational fears" or

ble senses of the ways an individual may be constrained can show that all the categories of "constraint" are not restrictions upon freedom. Dixon argues that constraints in the sense of "irrational fears, internalization of inhibiting norms" and constraints in the form of "limited conceptual framework" are not obstacles to individual liberty. The point is that "freedom from constraint" is an ambiguous term as such. We can extend its meaning to various contexts. Even "external constraints" can be extended to any extent. But once we go on adding "internal constraints" like the ones suggested by Taylor, there is no end to it. It may lead to the absurd extent of any psychological states. External constraints at least have some tangible standards to fall back upon. Led by similar kind of argument Dixon suggests that only the external constraints that follow from intentional actions or unintentional consequences of human agency, really put restrictions on human freedom.

Without probing further into the issue of whether freedom requires only the absence of constraints or some positive aspects like rational ordering of motivation etc., let us examine its possible relation with the social concept of equality. In other words do the two social goals (a) inconsistent or (b) compatible or (c) interdependent?

(a) A very crude version of either egalitarianism or libertarianism would regard that the two goals are inconsistent because if either equality or freedom is regarded as the social goal then from the principle of equality (or freedom whatever is the case) follows maxims which cannot be satisfied under the principle of liberty. For example from the principle of equality follows the maxim of equal distribution of wealth which is in contrast with the libertarian maxim of free-enterprise and holding of inherited property.

(b) The two goals can be compatible in a weaker sense if either the principle of equality or the principle of liberty takes a secondary position. This is a possible compatibility position with one single principle with a rider to accommodate the other as a negative requirement. For example, freedom can be the highest principle so long as it is *not* a hindrance to some other's similar freedom. Similarly we can have version of equality as a principle with a supplementary clause as in 'all are to be treated equally unless there are overriding grounds (such as merit or poverty) for some degree of inequality'.

(c) I am not arguing any more for these positions for there are plenty of arguments for them, some of which are already mentioned above. I shall try to examine instead the proposition (c) that equality and freedom are interdependent. The term "interdependent" may be used in a broad sense to mean both logical dependence and social desirability.

Two terms are logically independent if the meaning of them can be understood without referring to the other. The terms "Father" and "Mother" cannot be

---

compulsions, (ii) internalization of norms' that inhibits, (d) our limited conceptual framework, lack of education, false opinions etc. See his *Freedom and Equality*.

understood without reference to “son” or “daughter”. So these terms are not independent. Because the meaning of them depends on the meaning of other terms. “Tiger” or “Kind” on the contrary does *not* depend on any other term for their meaning. It is obvious that it is a very strong sense of “interdependent” and “equality” and “freedom” are not interdependent in the strictest sense. But the etymological sense apart, there are senses in which we use the term “interdependent”, particularly if we use it in social or political context. “Capital” and “labour” or “democracy” and “free market” are not in the strictest dictionary sense interdependent. That means the meaning of them do not depend on the meaning of the other, but clarification (in some cases definition) of one requires reference to the other. Otherwise the meaning remains ambiguous at best and incomplete at its worst. These types of terms used for expressing some concepts can be reasonably called interdependent.

The kind of contrast long associated with the concepts of equality and freedom treats them as totally independent concepts which stand for simultaneously unrealizable social ideals and there is a tendency to undermine any link between the two and the concept of equality is usually posed to stand against the concept of freedom. But clarification of either is not possible without reference to the other concept. In fact both libertarians and egalitarians contrast their principles and maxims with each other to clarify their position and also to justify the grounds for their own claim. We have already seen that egalitarians are not necessarily monsters who like to grab all private properties for a faceless social god. Neither is liberty a principle that allows indulgence to all degrees of destructive free activities for individual agents that are detrimental to rational ordering of priorities even if it concerns only the individual person and nobody else. Moreover in most (except in extreme cases) versions of these two principles a supplementary clause is attached to it where maxims contrary to the principle (and therefore conducive to the other ideal) are recommended.

The other possible meaning of interdependence is that these are either equally socially desirable or that the realization of one goal needs the other principle. Let us examine this requirement of social desirability with reference to a thesis known as equality thesis popularized by the feminists who used the phrase to refer to gender-equality. I shall argue with the help of some version of the equality thesis in this sense, which require freedom as a guiding principle.

The familiar version of the equality theory is proposed by Mill in his book *The Subjection of Women*[17] where he has meant to show that women should be on the same standing as men in law, government and society. It is generally believed that Mill holds the above view about women and also believes in the intellectual equality of the two sexes although he never argued clearly for either position.

---

[17] J. S. Mill, *The Subjection of Women*.

In a paper[18] David Stove made some scathing remarks about Mill's arguments (or rather lack of arguments) in *The Subjection of Women*. I am not going to judge whether Stove is justified or not in his tirade against Mill but shall try to show that even if Stove is right about Mill that the latter did not argue sufficiently for the equality theory (there are other vituperative comments which I leave aside) in this work - there is a possible move for Mill to establish equality thesis, namely, from the presumption of individual freedom (as proposed in *On Liberty*) as a prima facie good. Before doing that let us see why Stove thinks that Mill could not argue for the equality theory. Being a Utilitarian presumably Mill cannot argue for the equality theory from any supposed "abstract" or "natural" right to equality. Nor can he argue from the "principle of utility", because "he was emphatically not preaching to the converted here". His argument in this book was essentially an argument "from psychological and historical *facts*, or at least alleged facts". In short no known differences are found between two sexes, either of a kind or of degree.

So the legal and other "disabilities" woman are handicapped with, cannot be justified. It is obvious that what Mill needs here is an argument to show that there is no difference in the intellectual capacity of man and woman. The central mistake of Mill in Stove's view is his two propositions in this book namely: (i) The subjection of woman to man (and society) is unnatural and unjustifiable and (ii) the nature of woman is unknown in the sense that how much of it is natural and how much is due to external sociological condition is difficult to say, 'more research is needed'. The inconsistency of these two statements is quite evident. If we do not know the nature of woman how can we claim that their subjection is unnatural or unjustifiable? Moreover "more research" would not be politically correct or welcome to one who demands "more equality" for women. And what is so special about human beings that we cannot say that we know the difference between two sexes? Indeed we know a lot about the different sexes of animals and plants. And certainly there is a substantial amount of knowledge that we can claim to possess about men and women such as procreation and nurture of family is generally done by women while men take care of protection, procuring and some sort of domination over the folks. And much more. The only substantive point at issue here is that historical evidence of inequality in performance of women corresponds with their subservient position to men and society. But that sort of correlation does not show anything conclusively. Therefore mere historical evidences of subjection of women par se cannot establish that they are (naturally) equal to man. Given these three possible arguments (abstract or 'natural' right, utility principle and historical facts) - Mill cannot, unfortunately argue for the equality thesis.

---

[18] David Stove, 'The Subjection of John Stuart Mill', *Philosophy*, January 1993.

*Equality vis a vis freedom*

The historical facts of women's subjection, one can argue, show that they *were* treated unequally. So the difference does not show that man and woman are *unequal* in intelligence or any other talents. But at the same time there are no decisive reasons either to show that their capacity is *equal.* So Mill cannot argue in any possible way for holding the equality thesis. Granting that Mill could not argue in any way to support the equality thesis and granting also the grain of truth in Stove's diatribe that Mill *did not* argue for it, we should be charitable at least to accept that Mill *did* believe in the truth of the equality theory. But then how can he argue for it? None of the three possible ways, we have seen, can do that. One possible argument for the equality thesis could be an argument from the requirement of individual freedom. It is reasonable also to ascribe at least that motivation to Mill for his belief in the social equality of gender, particularly with reference to his principle of individual freedom as espoused in *On Liberty*. The argument can be articulated as follows: freedom is a prima facie value. It is desirable not for any other social good but for the sake of the individual herself. Following the principle of liberty (as in *On Liberty*) we can argue that individual freedom of speech and action is essential for the pursuit of an individual's own goal, one of the reasons for such requirement is fallibility of human reason. All opinions should be heard, all statements tolerated because they can be wholly true, partially true or partially false and wholly false.

Even in the last case there should be freedom of expression because contrasting with a false belief also is good for the development of our true belief. This is in nutshell the argument for freedom of expression from the fallibility of human reason. If freedom is a prima facie good, then all individuals of a society should enjoy it. But history of almost all societies can vouch for the fact that a large section of humanity was *actually constrained* by many external obstacles as well as internal factors (such as lack of information, fear, internalization of time honored norms etc.). Freedom of choice is an essential ingredient for human development. Unfreedom is disabling while freedom is necessary for individual pursuits. But if some people are *actually not free* at all, or not as free as many others then opportunities for at least them are not open in the same way as they are for others. And a non-egalitarian society is exactly that where everybody is not equally free. Here I allow that freedom can vary in degree. In other words, some people enjoy freedom more (for various complex reasons - social arrangements as well as individual traits) than many others[19]. Any upholder of equality theory -

---

[19] John Watkins argued that there may be different scales of freedom, one has a low degree of freedom if one is 'bogged down in otherness, thinking stale thoughts, or trapped in some lying ideology, and that one's freedom improves the more one succeeds, by dint of an inventiveness that one keeps under critical control, in transcending otherness and making a way of one's own that leaves the world a little richer than

be it equality between gender, among races or religions - would demand that everyone should be equally free so that all are equally exposed to options they can choose from. In this way we can argue for the equality theory from the presumption of prima facie value of individual freedom as Mill espoused even if it is granted for arguments sake that he did not argue this way in *The Subjection of Women*.

There is still one problem however that I may have to face in this argument from freedom. The problem is: in our argument it is supposed that freedom varies in *degrees* in all non-egalitarian societies, and so the degree of freedom one enjoys vary directly with the kind of egalitarian the society actually is. The most crude form of division would be into "haves and have-nots" which is vague, however, there can be finer shades of hierarchy. For example what about a hypothetical society where everybody is equal in the sense that no one is free? Logically speaking equality follows here from the presumption *not* of freedom but of absolute unfreedom. In other words an objection is possible that if equality follows both from the presumption of freedom and the negation of it - the situation is paradoxical or trivial. In other words if equality can follow both from freedom and unfreedom - we should not link up the notion of freedom with equality because the latter can follow from any principle and therefore, does not have any essential relation with freedom. These two are not only independent but opposite notions. If one is presumed, the negation also follows.

What could be the possible reply to such objection, which seems to be devastating? The objection will surely hold if either (a) 'It is possible to have a society where *all* are absolutely free' or (b) 'It is possible to have a society where *nobody* is free' - is true. First we do not need the truth of (a) because in this account we are not regarding freedom as an abstract right and (b) is presumably false. Let us see why it is false. For that we have to first distinguish between physical and social determinism.

Physical laws are inviolable in the sense that we are all equally determined by the physical processes so in that sense may be equally unfree. But social-political laws are man-made and of such nature that one can in principle violate their maxims although the particular kind of sanction attached to such violations depend on the kind of society the laws are operating in. The more coercive social policy goes with more stringent sanctions and more open a society is, more egalitarian are its maxims and less stringent are the sanctions. In a closed society like a religious state or dictatorial rule, amount of individual freedom is minimal and sanctions maximal. But even in the worst possible case of hierarchical rule

---

it would otherwise have been'. He is right that this difference is not elitist though in a way discriminatory for it does not grade individual as such but the various temporary lapse into stupor and different degrees of heteronomy or else he can rise high in 'states' of freedom an individual can be. It may be a scale of autonomy. Ibid. p. 255-56.

such as a totalitarian rule or dictatorship - most people may not have much freedom compared to the freedom or privileges enjoyed by the narrow group of ruling power or caucus around the dictator. If this kind of society is possible[20] (the worst case would be one where one person coercing all others) - even then we can see that *all* members of this society are not *equally* unfree, although most are. The small group of people within the ruling party or around it has different degree of freedom and in the equality scale these individuals are nearly equal to the highest power depending on the options and opportunities open to them.

The ground of such hierarchy of options however is not rational; it is based on rather the amount of allegiance of these people in the implementation of the coercive policy perpetrated by the ruling power. So it is really a trickle of privileges flown down to them. One may of course contest whether that kind of options are really free because these are merely prices (or bribes) paid for these people's prior commitment to unfreedom preached or perpetrated by the dictator. And if any of the caucus withdraws support from that policy, he would probably lose that little freedom (say freedom of choosing his house or enjoying the freedom of financial abundance which can further his various other pursuits). Indeed the stark example above is the worst kind, and we can also think of cases where hierarchy is the least. Nevertheless it is still not true that 'all are equally unfree'. Because in the worst case at least the dictator is *free* to invoke rules and sanctions to impose coercion and at most there is a hierarchy enjoying degree of freedom corresponding to their nearness to the ruling power. So in neither case we can say 'nobody is free' (is true) as we can say in case of physical law. Moreover this case also illustrates why rule by coercion, even if it brings equality among most (not unlike the equality among slaves) cannot be a social ideal. Egalitarian principle only cannot justify why equality is not enough. To make it socially desirable we need the principle of liberty as a prima facie requirement.

Another unfamiliar version of the equality theory can be formulated[21] as follows: Women and men are *equally* human. For various historical and sociological reasons women are perceived as something which they were made into. One important reason specially for a unfree nation (such as in India under the

---

[20] Of course one has to allow that certain special emergency conditions like war, famine, natural disasters etc. call for temporary suspension of many individual liberties. But indefinite continuation of such 'emergency' period however would amount to deprivation of freedom.

[21] In a letter to the poet Sukhalata Rao, the Bengali Nobel laureate poet and philosopher Rabindranath Tagore wrote about the issue of man-woman equality. Tagore's letter is a response to Sukhalata's article in Bengali 'Naree' (Woman) which the great poet finds a balanced and good exposition. Tagore's letter contains implicit arguments for the equality thesis. It was later published in Bengali with the title '*NareerManushatta* (*The Humanity of Woman*)'. This was in the 1930s although the letter was published in 1997 in *Desh* (a Bengali literary magazine). The concept of woman as a social construction has become popular after Simone de Beauovoir's *Le Deuxième sexe* (1949).

British rule) - where men were not free in any other sphere except their own home where only they could reign supreme. To keep their supremacy and ego-satisfaction intact women in the family were treated as subordinates. Servile roles were ascribed to them, sanctified by the approval of religious texts and rituals. Marriage is such an important institutional license. The most dangerous threat to freedom is that the norms sanctified by social approval were gradually internalized by women in such a degree that not even external agencies are needed for their compliance. Women on their own cannot breakaway from the social prisons they were captives in. In Rabindranath Tagore's[22] language 'the three fourth of what we call a woman is constructed'. And, we all know that the constructed image is of one who is inferior in all kinds of talent in addition to physical prowess and whose role is to satisfy all sorts of demands made by men around her and follow the rules of society arbitrarily made only for her obeisance.

Indeed in a closed society like a country ruled by a colonial authority, nobody is as free as one in an independent society of a free nation. But within the circumscribed framework, women are treated unequally. The reason behind this inequality lies in a misunderstanding of the concept of humanity. The Enlightenment ideal (which influenced both Mill and Tagore) sets us against any other value than reason. If rationality prevails then it is easy to realize that so far as humanity is concerned men and women are equal, both are human beings with similar capacity of reason and emotion, potential of talents and abilities. Biological, psychological differences notwithstanding, these two species of the same class "human" - are equal. This equality will be evident only if we realize and detect the constructionist elements in the concept of woman. The feminists in general emphasize the constructionist aspect of the concept of a woman and made it an issue to fight for the true image of a woman as equal to man in all social political roles including intellectual ones. "Empowerment of Woman" is actually a cry for representation of woman as a free person equal to her male counterpart.

Tagore's standpoint despite its arguments being often mixed up with reference to the prevailing political condition of the time is quite clear. It indicates that the most important point, which is usually missed in our concept of woman, is the realization that the concept is first one of an individual human being just as the concept of man is that of an individual human being first. This insight is essentially true to the Enlightenment notion of "Man." In that sense Tagore believed that if in any society half of the people are treated unequally, deprived of many opportunities and knowledge - that kind of society can never progress - whatever development is achieved by the other half. Because progress there would be lopsided and no society can go forward without over- all development. Of course he does not mean women only but all other underprivileged classes in

[22] Ibid. p. 37.

the lower rungs of social hierarchy. Elsewhere he says 'the persons you leave behind will pull you backward' and so inequality can eventually be suicidal.

Now a question arises at this juncture: Why inequality is not prima facie desirable? Why can't one argue that some sort of hierarchy is needed for social stability? The reply could be as follows: if women (or any other underprivileged class) are treated as *not* equal then it follows that there are others who are *more* privileged. If so then we can make a distinction at least of degree of deprivation of rights. And if woman is first a human being then no rights can be denied of her which is being enjoyed by the other counterpart of human being, i.e., man. In other words, if freedom is a prima facie value for human being then that value cannot be denied of women as well as men. And once we grant that both men and women are human beings we have also to admit that freedom is a prima facie value for both. So women should be treated equally regarding their rights notwithstanding biological and other natural differences they possess to be called woman.

## *Concluding Remarks*

The tension between the two concepts, equality and freedom often leads to opposite social ideals. It has been attempted here that the two philosophical concepts are *not* incompatible. Before going into the compatibility and interdependence issue, we can consider different positions that allow different status to these concepts. John Rawls' "Just society" and the principle of liberty are examined where *inequality is not* regarded as prima facie wrong; on the contrary some kind of inequality is justified to give supremacy to the principle of individual freedom. Some extreme form of egalitarianism however can be positioned against the Rawlsian version. It asserts that equality is preferable in its own right; and the existing inequalities stand in the way of human self-realization and autonomy. This is again a utopian presentation of a co-operative sort of society where opportunities and affluence are utilized to the maximum benefit of the people. Kai Neilsen has envisaged this kind of society where "justice-based equality" is the guiding principle. In both these versions there are internal and empirical problems. The internal problems are that of complexity of nature and the empirical problems are mostly difficulties related to implementation. These difficulties aside, the crucial question here would be (given that some sort of inequality is inevitable): is equality *per se* a social good or guiding principle? Is it so sacrosanct that individual freedom can be a secondary principle?

One can of course argue that equality does not have any intrinsic value. It can also be pointed out that the assumptions of egalitarianism are either false or unwarranted, because this view presumes that *all existing* principles of distribution are wrong. There are further problems if we assume this, and formal framing of a general principle of equality is also very difficult. Because it invokes an abstract rule that appeals to our intuition of natural justice. But egalitarians still have to face many difficulties regarding private holdings and entitlement and clarify their position on this issue, which seems to be counterintuitive to many.

Freedom and equality are usually thought to be inconsistent principles since the maxims that follow from them seem to be contradictory. The notion of freedom can be understood in two senses, freedom is a requirement for principle of equality and therefore the two notions are not incompatible. On the other hand they are interdependent, not in any strictly logical sense, i.e., understanding the meaning of one is possible not only with the help of the other - but in the sense that the clarification (in some case definition) of one requires reference to the other. Interdependence can also mean that these are either equally desirable social ideals or that the realization of one goal needs the other principle.

Two versions of the equality thesis namely Mill's well-known thesis and the not so known thesis of Tagore can both show that it is possible to argue for gender-equality (as well as equality among people of other classes) only from the assumption of the principle of individual freedom.

# *CHAPTER III*

# TYRANNY IN DISGUISE: EXPERT KNOWLEDGE - A NEW FORM OF PATERNALISM

'...The appeal to experts ... is
an intellectual fashion ... and the
appeal to authorities is the death of
knowledge and ... the growth of
knowledge depends entirely upondisagreements.'
Karl Popper, 'Author's Note, 1993', *The Myth of the Framework*, 1994

## *I*

The most complicated component of social/political freedom is the capacity to make a conscious choice through knowledge and intelligence. In a sense animals are freer than humans are in so far as they pursue their goals instinctively. But human decisions are made by conscious and deliberate choice. Apart from other impediments like natural and political ones, human individuals have another constraint to overcome to make a free and deliberate choice, namely *tradition*. The long history of human civilization has an overwhelming pressure on individuals in the form of complicated customs, creeds and value systems which are positive determinants to influence free decisions. These informal restraints are more powerful than formal restraints like judicial and political laws and often more powerful than formal restraints like judicial and political laws and more often than not the latter kind are based on the former. Tradition in different forms and degrees has sway over an individual choice depending on the kind of society it is made in. The older the civilization the more rigid and complex its codes are with stronger grip over the individuals. The internal constraints a tradition imposes upon individuals are like sense of dutifulness, identity problem feeling of sin or righteousness, which are positive constraints for a free choice.

*Tradition* and reason seem to be two conflicting forces that an individual faces to make a conscious choice. Of course a choice made in compliance with the tradition may be rational, but then it does not require a conscious effort to give the reasons in favour of it. But an option against the current of tradition needs great efforts to rationalize it. An option blessed with the approval of tradition does not require that great effort for it is going with the tide and not *against* it. In a totalitarian system or even in a society of long tradition of rigid conformity and insular social ambience one may *feel* free for lack of access to information or opportunity for other possible choices. But *being* free is much clearer than feeling free and measurable by tangible social norms.

Mill and Tocqueville classically present the repressing influence of custom or tradition over individuals and the debilitating effect they have over free choice as 'tyranny of the majority'. This kind of tyranny as they rightly put is more enslaving than political tyranny because one can rise against political power, but custom and public opinion work from within to resist any effective protest.

The growing sociological interest in tradition and its role in culture tend to breed a kind of conservatism in the recognition and respect for various traditions. This pluralistic view of culture and tradition is based on an assumption that reason itself is contextual, i.e. contingent upon the tradition itself. Therefore the correct evaluation of a tradition is difficult because we cannot judge it by any norm outside it. Thus there seems to be a tension between two approaches; one, tradition is against *reasoning*, the latter taken in a very circumscribed sense, that of scientific/logical reason. Two, tradition cannot be judged according to any standard, it is like an inevitable process to be sociologically explained, but not rationally adjudged.

But if a tradition is, or at least claims to be based on reason it will be much more formidable an obstacle for a free choice than any other traditions. The pressure of the pre-scientific traditions was at least not so powerful as compared to the tradition of science and technology.[1] The motto of the new tradition is 'knowledge is power' which is initiated by Francis Bacon's euphoric acclamation of science and its victory over nature. His naive trust on the power of (scientific) knowledge was based on the "correct" method of induction, which only can ensure truth about nature. Interpretation, use of hypothetical methods or any flight of imagination was unscientific to him. Only the "inductive machine" can churn out truth and nothing but truth to anyone who is unpolluted by any extant theory or "prejudice" and only guided by observed facts through the looking glass of the inductive method.

Indeed the romantic ideal of science as an objective rational epitome of truth is no more entertained by any mature theory of science. Neither is it regarded any more that a theory can be logically derived from some elementary statements. The battle among rival views about the goals of science and its methodology is still on. But nevertheless science as a tradition is the strongest tradition ever known since it is supposed to be based on reason that encompasses all traditions and thus has a sweeping power over everything else. In other words scientific theories (beliefs) reign over all other kinds of theories for their universal validity irrespective of other differences of the subjects under inspection.

[1] Herbert Marcuse in *One Dimensional Man*, envisages little freedom in the traditions and organizations of modern industrialized society. The system creates its own needs and ways of fulfillment in such a way that we are witnessing the emergence of 'one dimensional man' - creatures incapable of thought or action in any other dimension than the established order.

This omnipotent picture of scientific theories however has a paradoxical element. In one way science made individual lives free from many old drudgeries and improved the quality of life to a tremendous extent but on the other hand it creates a condition where individual freedom is always at jeopardy because of constant dependence on the powerful influence of scientific knowledge or at least the claim to scientific knowledge.

An adequate analysis of the problem of a possible tyrannical domination of scientific knowledge over individual action (no pretension is made here to suggest an adequate solution) requires first to examine the nature of scientific reason in general and expert knowledge in particular. And after examining the arguments to show that such domination is detrimental to individual freedom, and *how* it is detrimental, we need to indicate how common-sense view and ordinary reason are undermined. But the question still remains why despite all its imperfections and apparently paternalistic attitude science is until now the most tolerant and powerful tradition. After all science is whatever else it is (to borrow Susan Haack's phrase), 'a rich powerful and prestigious institution'. But first the problem of a possible tyrannical domination over individual choice.

As mentioned before Marcuse's basic arguments against the "new form of control" are as follows: individual freedom is being rapidly suppressed in all areas of our life because of the demands of technological-industrial production. Moreover, this technological tyranny has become almost invisible to most members of our industrial civilization because the propaganda of the mass media is so tremendously successful in making us believe that we are actually free. Marcuse is pessimistic about the possibilities of changing this course of development. He argues further that paradoxical as it may seem the very success of our industrial civilization makes both freedom and effective opposition to tyranny increasingly impossible. In our society today, technological controls and the growing technological tyranny are presented as the embodiment of reason, and therefore an opposition is seen as irrational and unrealistic. This is especially true, when technological control is presented in the guise of expert knowledge. Marcuse's main contention is that the radical empiricism and experimental basis of modern scientific thought, which he believes has replaced all other criteria of knowledge, defines all forms of knowledge in such a way that one cannot be critical of the status quo. The basic forms of knowledge make it irrational and absurd to think in terms of anything other than what already exists. This system of thought and its social system absorb all criticisms and make all opposition into a support for itself.

Marcuse's main point about what he calls one-dimensionality of man is basically well founded. Indeed an industrialized society creates new needs and specifies the ways to fulfill them so that individuals gradually internalize those norms so completely that they do not realize that these are imposed by the society and not their free choices. But everybody will not share the epistemological pessimism this realization indicates. Of course expert knowledge has a powerful social role to play, but that does not necessarily mean that it cannot be a suspect. There are methods of checking and cross- checking *within* the scientific practices that is meant to keep a guard on illegitimate exercise of specialized knowledge. I shall come back to this later in the discussion.

There is another possible argument from the complex social relations resulting from the growth of the mass society.[2] Technological mode of production in this view makes centralized social planning absolutely necessary. The question is whether the social planning is not itself the result of rationally planned action, if it is done in a piecemeal fashion, then very likely the form of social planning will be totalitarian. If the planning is done for purpose of upholding our individual freedom and supporting our democratic institution, it can become a fundamental force supporting democracy and leading to greater freedom. So social planning and democracy is not basically opposed to each other.

The use of scientific knowledge to gain this objective and reliable understanding is needed to make rational plans for such highly complex society. Mannheim in contrast to Marcuse is very optimistic to think that wise and courageous moves would help plan for democracy and individual freedom. Modern technological knowledge can be judiciously utilized to greater individual freedom and greater economic affluence for all. The technological society like our modern knowledge-based society can either be a society of wealth and freedom or can turn into a terribly efficient tyranny worse than any one known so far. After all techniques are neither good nor bad, their use is. Mannheim called for a 'militant democracy' for the realization of these hopes and at the same time to guarantee that the technicist projection remains a nightmare forever.

That our present technological knowledge-based society is a mass society different from earlier forms of society is a correct insight. And the optimistic hope that a watchful conscientious democracy can ensure a right application of knowledge-based technique is stimulating. Indeed social planning is not necessarily an evil and what Popper calls "piecemeal engineering" or gradual social change rather than drastic total change can be more effective as it works through individuals rather than a social power from above. But the point is, knowledge gives power and technology and expert knowledge can often be misused for "social cause". It is not possible in principle for any democratic institution however "militant" it is to monitor all decisions made. There is an inevitable elitist or exclusive element involved in decisions concerning scientific/technological affairs. No democratic force can fully censor that. Ironically a totalitarian system can watch over (and even control) scientific decisions, but that is for a so-called social cause or an ideology rather than for any democratic motive. Moreover interfering with scientific decision would be interfering with scientific freedom. Thus in neither way it is possi-

[2] K. Mannheim, *Diagnosis of Our Time: Wartime Essays of a Sociologist.*

ble to have access to scientific decisions without hampering freedom of someone.

## II

The most serious and irreverent attack upon the authority of science and practicing scientists has been launched by Feyerabend. Along with his rejection of the rationalist image of science as one with truth and objectivity as a goal by referring amply to history of science, he argues that rationality or truth was never the methodological criterion for theory acceptance. There are many extra-logical factors like propaganda, conceit and clique that made one theory winning over another. Scientific community is a closed insular community basking in the glory of their high status in the society. There is no ground, he thinks, to believe that scientific reason is better than ordinary reason. He points out in several works that the latter is undermined unnecessarily by the so - called expert knowledge and its acclaimed superiority.[3] Feyerabend saw prevalence of science as a threat to democracy and individual freedom. Before going into the issue of how it is a threat and what kind of remedy (if there is any) can sustain individual freedom as well as autonomy of scientific research programme let us turn to the question of science being only *one* of various traditions and how the practicing scientists often overstep the boundary in total lack of awareness about the existence of other possible traditions.

For many sociological historical factors western scientific tradition now is the most powerful tradition. And to borrow Feyerabend's phrase the scientists now are the high-status priests and science is the new religion. I said 'for sociological historical factors' only to show that there is no intrinsic worth for it to win if the historical accidents (?) like colonization, western suppression of the vanquished culture, imposition of the western way of life etc. would not have taken place. By this I do not want to convey that modern science has not achieved anything, in fact it's tremendous achievement in the last three hundred years has been so astounding that we seem to forget its historical nature. I will come back to the point of how this oversight generated many misconceptions that are entrenched in our modern culture and even use of language. In a brilliant essay[4] Feyerabend has argued that the conflict between Galileo and the Church

---

[3] 'The Strange case of Astrology' in *Science in a Free Society*, P. Feyerabend. By citing a paper signed by '186 leading scientists' including 18 Nobel laureates disparagingly condemned the growing acceptance of astrology in various parts of the world -- Feyerabend has argued that it shows 'the extent to which scientists are prepared to assert their authority even in areas in which they have no knowledge whatsoever'. See also *Farewell to Reason* particularly chap. 6 'Trivializing Knowledge': and chap. 9 'Galileo and the Tyranny of Truth'.

[4] Feyerabend, *Farewell to Reason* Chap. 9, 'Galileo and the Tyranny of the Truth'.

is really a conflict between *traditions*. The difference is the difference of traditions concerning the role of *experts* in society.

One tradition regards an expert as the final authority on the use and interpretations of expert views and expert procedures. The other subjects the pronouncements of experts to a higher court either of super experts (this was Plato's view) or of all citizens - this is perhaps recommended by Protagorus. The opposition between Galileo and the Church was analogous to the opposition between the first and the second. Galileo was an expert in the domain of mathematics and astronomy. He did not only want *freedom* to express his view to public but also wanted to impose it on others as the only correct method to have access to truth and reality. The position of the Church was like the second position in its Platonic form. According to the first, traditional society must adapt to knowledge in the shape presented by the scientists. This is the tradition defended by Galileo according to which the principles of the Church (or any other tradition) are made consistent with the findings of the natural sciences. This is Bellarmino's (representing the Church position) stand except that expert knowledge from a special and narrow domain has now replaced the wider and more humanitarian point of view of seventeenth century Catholicism.

According to the second tradition scientific knowledge is too specialized and connected with too narrow a vision of the world to be taken over by society. It must be judged from a wider point of view that includes human concerns and the values flowing there from and its claims to reality must be modified so that they agree with these values. Feyerabend thinks that the Church with its wider expert knowledge and the 'eminently human document the Bible', 'had and still has a tremendous advantage over the principles of an abstract rationalism'. Of course the noble sentiments did not always prevail and some Church directives were merely exercise of power -- but there were nevertheless some predecessors of modern attempts to counteract the dehumanizing tendencies of so-called scientific objectivism with the help of some human elements. If we realize this concrete humanistic aspect of our knowledge about the world the claim of some elementary particle physics to have found the ultimate constituents of everything have to be rejected and replaced by a more "instrumentalistic" position. (I shall argue later that it is difficult for scientific realism to dislodge the claim of common sense that ordinary objects like apples and oranges are real). Scientific theories are not about reality; they are about predictions of phenomena in a reality determined independently of their efforts. At the time of Galileo the second tradition of humanitarian nature was the one defended by the Church. Feyerabend seems to endorse the kind of authority Bellarmine possessed with "wisdom and grace" against the abstract principles of Galileo (which according to him involved lot of problems including its basic physics being questionable). He finds it surprising that the Church in the present time has never contested any of the scientific claims seriously nor of scientific ideology as a whole. So there are no other competitors in the field and scientific tradition wins as if it is the only tradition.

I do not share Feyerabend's vituperative diatribe against science as a tradition as well as its practitioners the so called "experts" of specialized knowledge to that extent but nevertheless share his basic philosophical insight that empirical success (if we ignore the many lapses and failures) of scientific research should not be counted as its sheer logical or abstract strength. The dazzle of achievement should not blur the actual historical perspective in which modern western scientific research tradition should be placed.[5] And to do that the first question to be raised (and answered if possible) is, why then scientific tradition is regarded as the most powerful tradition or whether it is powerful and what role we should ascribe to the practitioners namely the experts in specialized areas.

I shall first consider the possible reasons for science and its practitioners playing the role of authority in society and then examine the alleged tyrannical element in such authoritative role. The most important standard argument to vindicate the superiority of science (and its men) is drawn from the methodological point of view. Philosophers of science may differ as to what should be the criterion, but they are unanimous in voicing that scientific research follows certain set of rules both for theory construction and theory evaluation which are if not strictly computational, at least based on rational ground. In other words scientific method is the touchstone which makes the research programme different and superior to other traditions. The older positivist's strict logical criterion was too abstract and narrow and therefore soon abandoned. But the hypothetico-deductive model of Popper and Hempel also (even with the falsifiability/testability criterion) is inadequate to explain many phenomena without some ad hoc or auxiliary hypotheses, that means by some ad hoc adjustments with the initial conditions and the theory in question.

Moreover examples abound in history of science that there are no single procedure or scientific method. Every theory, every project has to innovate or evolve certain methods, therefore, these are to be evaluated by its own standards and merits. The idea of a universal and stable *method* and also the idea of a universal standard of *rationality* have become a suspect. As Kuhn's celebrated attempt has shown, normal scientists do not abandon a theory as soon as they detect an anomaly; on the contrary they go on trying to fit it ('puzzle solving') to the 'paradigm' or the dominant theory they are working under. So the picture of a normal research programme is not one of *rational criticism* but fitting data to the theory. Lack of immediate success is not necessarily a fault of the method but

[5] Prajit K. Basu, 'Reappraisal of the Civilizational Question: Why the Scientific Revolution did not take place in China or India', *Sandhan, Vol. ii, No. 2, July-December 2002* argues that it *ought* to have taken place in these countries in an attempt to answer the *grand* question raised by J. Needham in *The Grand Titration-Science and Society in East and West*, London, 1979, p. 16.

indicative of complexity of the problem and needs more rigorous research in the area (Cancer research is such an area) and might require change in the research strategies or even uniting one research programme with another or subsuming one under another. Einstein has remarkably done that by changing the frame of reference when the prevailing framework (Newtonian) seems to be inadequate although the latter stayed on as an approximation. Incidentally Feyerabend's invective includes "There is no reason why the research programme *science* should not be subsumed under the research programme *free society* and competencies changed and redefined accordingly".[6] This is an extreme position and involves further problems that we will see later.

Another well-known and well-accepted argument for the superiority thesis is that so far the empirical success of science is uncontestable. Yes, no one can deny that scientific research in the last two hundred years particularly, has made tremendous contribution to our understanding of the world as well as of practical convenience to us. Here I am ignoring the minor lapses and stories of failure. But as I said before the success story also coincides with some other stories from history and politics. Expansion of Western power to other countries full of natural resources as well as human resources went along with the achievements. Moreover the ancient traditions were either demolished or ignored and not had a chance to compete with the growing new scientific activities. I agree with Feyerabend in this point. So it is difficult to say whether other traditions (having had a fair chance) would not have been similarly successful. I am not by any means underestimating the *actual* accomplishments of scientific research but only asking any euphoric admirer of modern science to be aware of its contingent nature and the *possibility* of other kinds of achievements if actual history *were* different. Moreover science as just a pursuit of truth could not have been so powerful unaided by the technological revolution it brought. Technology brought affluence and power which again makes it possible to continue further expensive and efficient research and so on.

Another corollary of this argument exploits the sui generis nature of science. That means science is a new phenomenon, an autonomous emergent event that is entirely different from all earlier traditions.[7] Even events or periods are divided as "scientific" and "pre-scientific". It assumes either that there were no sciences before that period (say seventeenth century) or if there were some tradition, that is not science if anything, or both. What to make of this argument? If by "scientific" one means modern scientific then of course there were no "modern" science in ancient time or before the stipulated period 'modern' and it is an unnecessary semantic exercise. But if by science one means knowledge or understanding of the world and a way or method to cope with the problems related to it then of course there were science in all stages of human civilization. And although

[6] P. Feyerabend, *Science in a Free Society*, p. 100.

[7] Both Popper and Feyerabend deny this.

modern western science tradition is a new phenomenon no one with a sense of historical perspective would deny that ancient systems of science, even myths, legends, tribal religion and their customs are predecessors of many modern scientific theories.[8] The ancient sun worshippers, Popper thinks led the way eventually to the heliocentric theory of Copernicus. The ancient Egyptian's problems with the annual flooding of the Nile make them measure the extent of the floodwater and thus gradually geometry is developed. So if we assume that everything before 17th century was "unscientific"– even then the role of such unscientific traditions cannot be denied as the precursor of the scientific era. And again it is not unreasonable to assume that there actually *were* systems of scientific thought in ancient cultures such as the Egyptian, Indian,[9] Chinese and many others.

The development of scientific theories again has a history of return after a defeat by a rival theory. The atomic theory has such a history. The ancient Ionian atomism to explain the phenomena of motion or Indian atomism to explain the phenomena of motion or Indian atomism (Vaishesika) were later defeated. The sophisticated physics of Aristotle did not require atomism and rejected it. The scientific revolution again saw the rise of atomism which again is set back by continuity theory. Late nineteenth century witnessed its return but again it is pushed back by the complementarity (of particles and waves) thesis. Not only theories but also methods and ideologies can come and go. Knowledge founded on logic and mathematics was supplemented by empirical methods (with a change in the ontology). Descartes, Leibnitz and Galileo stressed on mathematical methods, which again was replaced by radical empiricism of the powerful Copenhagen tradition.

There is another point first raised by Feyerabend that even if there were competition between ancient western science and philosophy 'there was never any fair competition between the entire complex of ideas and myths, religions, procedures of non-Western societies'. He also adds quite acidly 'These myths, these religions, these procedures have disappeared or deteriorated not because science was better but *because the apostles of science were the more determined conquerors*, because they *materially suppressed* the bearers of alternative cultures.'[10] The invectives aside, the main contention can be formulated as follows: the actual victory of modern scientific tradition does not necessarily establish its superiority over other traditions because of many institutional and contingent factors which were favourable for the vigorous research opportunities enjoyed by science and denied of the other traditions. I shall close this section with an

[8] The Jesuits in the 17th Century while in the Amazons learnt from the locals to extract quinine from the barks of Cinchona trees to fight malaria. The knowledge was absorbed in the general body of science much later.

[9] See, B. N. Seal, *Positive Sciences of the Ancient Hindus*.

[10] *Science in a Free Society*, p. 102.

example[11] of what could have happened if the institutional factors were removed or not present.

The history of traditional medicine in China can show us how important the social and political factors are in the excellence of a scientific practice. Till the 19th century China was one of the few countries that was free of Western intellectual domination. But then early in this century old traditions and the restrictions give way to be impressed by the material and intellectual superiority of the West. Science made an inroad to modern China and pushed back the traditional elements like 'herbal medicine, acupuncture, moxibustion, the yin/yang duality, the theory of the Chi were ridiculed and removed from schools and hospitals'. Western medicine was upheld as the only system. But in the fifties the party realized that they need more political control over scientists and ordered traditional medicine back to the universities and hospitals. So the two rival systems stayed on. One now discovered that traditional method of diagnosis and therapy are far superior (and less expensive) to those of Western scientific medicine. Another example to show that the division of pre-scientific and (Western) scientific is arbitrary and that the modern scientific tradition (say from 17th century onwards) is not a new phenomenon but a continuation of a long tradition (of often better understanding of nature and achievements taking the amount of obstacles to research into account).

In the latest Needham volume of *Science and Civilization* there is not only account of biological technology of traditional China, but also ample evidence of transfer of technology from India. "Despite some exceptions, it is generally agreed that the art of making crystallized sugar was not originally a Chinese invention, but came from outside, most probably from India. Chinese scholars of the Sung, Japanese scholars of the Edo period and finally The Europeans, Ritter, Von Lippman, Deerr and others all subscribed to the idea that a technological mission sent to India in the mid +7th century by Thai-Tsung, the second Emperor of the Thang dynasty introduced the art to China... the Chinese were already boiling low-grade amorphous sugar by about +4000".[12]Most scholars cite the following passages concerning Magadha in the *Hsin Thang Shu* (p. 6239) which was compiled in +1061: "... Thai Tsung despatched an envoy to obtain the method of boiling sugar and instructed that sugarcane be submitted from Yang Chou. The juice was (extracted by) pressing (in a mortar) as if (preparing) a dosage of (medicine). It far surpassed the Western regions in colour and flavour".

Ancient Indian medical literature of the Charaka Samhita (AD 100) and the Susruta Samhita (AD 300) mentioned sugarcane varieties or clones. In the 9th

---

[11] Ibid.; also, J. Needham, *Science and Civilization in China*, in six volumes.

[12] J. Needham (ed.), *Science and Civilization in China,* vol. 6: Biology and Biological Technology. No. 3, p. 368: (1) *History of Chinese Sugarcane Technology* (ed. C. Daniels). (2) *History of Forestry in China* (ed. N. K. Menzies).

Century the Muslim botanist Abu Hanifa Ad-Dinawari writing on the plants of east Africa mentions white, yellow and black clones. In this context Daniels remarks... 'Indian Buddhists did have a potentially better understanding of these matters than Europeans...'. 'The Buddha discusses right views. He says that just as a nimb seed or a bitter pumpkin when planted, even under good conditions gives bitterness and unpleasant taste. Similarly sugarcane, rice or grapes planted under varying conditions always give products of sweetness, pleasantness and delicious flavour. In both cases ... the cause of these is in the nature of the seed". He further comments 'It took the Western world until the 20th century to reach the clear concept of genotype and environment implicit in this text'.[13]

## *III*

Before moving on to the question of scientific/expert knowledge being tyrannical let us explore the possible grounds for doubt in the omnipotence status usually attached to science and its practitioners. Indeed there is fairly large number of grounds to suspect so. Below are some of the possible reasons for such, but the list is by no means exhaustive.

First: Choice of problem areas for research and the priority scale is often arbitrary or determined by extra-scientific factors. Even other methods of research such as traditional methods or belonging to other traditions are never given a central place. Goal of research in general seems to be not so much alleviation of human suffering as for concentration of power--mostly destructive power. The choice of the problems is crucial for the perennial problem of limited fund and human resources.

Second: Even within the domain of scientific research high-prestige, low-prestige areas are marked arbitrarily. For example, why physics (even there are hierarchy within it) and not human physiology enjoys a high prestige status? After all human body is more interesting and relevant for our concern than the macro or micro world of external entities outside it. Other life sciences usually are low priority (and hence lower status) areas.

Third: It has been felt by many that it is time we realize that scientists are not free seekers of truth, but rather after matching experiments with a dominant theory. If there are apparent anomalies to a theory, these are regarded as aberrations or personal mistakes and not refutation of that theory. So if scientific community itself is not a free enterprise how can there activities be conducive to individual freedom?

[13] Ibid., pp. 166-167.

Fourth: Feminists[14] also bring the charge of sexist bias to science and its priests. They point out that this bias is ubiquitous throughout the tradition-choice of problems (women's problems are not given enough importance), the position and number of women in the scientific-profession--all are indicative of the inferior status of them in the male dominated world of science. This under-representation is not due to any basic reason of lack of merit but due to male bias. In fact the general attitude of science as an adventurous endeavour to explore and ruthlessly exploiting nature is similar to male domination over women. The rhetoric usually associated with science "conquering" or "enslaving" nature reeks of gender bias. Examples of experiments with weaker sections of nature (animals, poorer countries, women etc.), denudation and pollution of natural environment etc. speak of stereotype male image of domination over and control of nature in its many facets. The "spare parts" of people of poorer countries, scientific experiments with weapons there and radiation-exposed medicine and other articles with people in these countries are only a few cases to show the dominating nature of science over weaker people. Even the language of science, some feminists claim, reeks of male prejudice of dominance. The search for 'Master molecule' etc. is indicative of such bias.

Fifth: Professionalisation of scientific tradition makes it more of a lucrative career than search for truth as envisaged by the classical romantic ideal of science. Fund, resources, awards, tenure etc. made the picture muddier than ever. It is a general knowledge now that lobbying is no less effective if not more than intrinsic merit in bringing official honour to a scientific professional.

Sixth: In the past science and its high status "reason" often supported racial and other kind of prejudices by allegedly scientific research in the past. For example there were "scientific" research in the past to show that blacks and Jews were regarded as (intellectually) of inferior brain capacities by so-called scientific reasons.

Seventh: As pointed out by many philosophers including Feyerabend, the tall claim of scientific/expert knowledge is unwarranted because often scientific theories are shown to be false. The history of science indicates that it is possible to explain the same phenomenon by more than one theory. Therefore it is not reasonable to have any unnecessary faith on expert knowledge because all knowledge is fallible.

One can also point out that the traditional image of science as the epitome of logic-rationality-objectivity is eroded from within the scientific tradition itself by the principle of indeterminism. Moreover methodological strategies such as seeking a "second opinion" in case of medical diagnosis, use of placebos, dou-

[14] *Feminist Approaches to Science*, ed., E. Fox-Keller, contains several papers dealing with different aspects of gender-bias in scientific practice. See also H. Seigel, 'Genderized Cognitive Perspectives and the Redefinition of Philosophy of Education', *Teachers' College Report* 85, 1983.

ble-blind tests etc. are indicative that the man of science himself does not have unflinching faith in the sacrosanct status of scientific method. So the anxiety is not groundless that optimal reliance of layperson on the sanctity of scientific knowledge as gospel truth rendered the tradition immune to doubt and criticism. I shall come back particularly to this point later because there is lot of complex factors and therefore a possibility of confusion in this connection.

The basic point emerging from these grounds to suspect the infallible status of scientific knowledge is as follows: There is no unquestionable reason to regard science in general and the scientific experts in particular as the neutral arbiter of truth and rationality. The norms and standards set by science and its champions are arbitrary if not irrational. For example if astrology is unscientific then what about psychological tests that replaced logical evaluation (like astrology as alleged by the scientists) of a person by some tests (which again are not beyond doubts) and methods not falsifiable in nature? Alternative medical systems again are not sufficiently explored at par with the standard medical science despite the innumerable failures and the enormous expense the latter involves. There are reasons to believe that such indifference is due to extra-scientific factors. One clear case is the emphasis given in therapy area including expensive surgery and diagnosis processes than the area of preventive medicines. Ancient traditions emphasized on the latter, and that saves lot of trauma, suffering and expenditure. But therapeutic medicines surgical and diagnostic accessories are matters of big business and lot of money and interest are involved there. So it is not easy to dislodge their priority. Alternative medical traditions are desirable for another good reason: there is least interference with human body *before* or *during* the treatment. Feyerabend is at least right in pointing out that the dignity of human body is preserved and respected in other traditions. He mentions the Chinese only but it is applicable also to Homeopathy, Ayurvedic or Unani systems. If empirical success is not guaranteed by medical science, then unflinching pain, injury, loss of organs or limbs seems to be unnecessary infringement of human freedom.

Indeed, there are cases where alternative traditions are sought but that happens only after the scientific treatment seems to fail. But otherwise and in other fields the scientific opinions reign supreme and work in a paternalistic mode when encountered with ordinary culture/reason. In a sense this move is tyrannical because expert knowledge seems to be autocratic and individual right to information is often denied or supplied with insufficient and biased information. Expert system[15] itself has become a suspect. Consciously or unconsciously ideological commitment can prejudice or at least influence empirical investigations.

---

[15] See W.E. Stumpf, 'Peer Review', *Science,* 22 February 1980, pp. 822-823, for such accounts. Cyril Burt, a psychologist, had been accused of fabricating parts of his twin studies which had been substantiated by his biographer and successor L. S. Hearnshaw. See 'Preface' to the latter, Cyril Burt, *Psychologist.*

Cancer research reports are often evaluated on the basis of *which* group has undertaken the research.

Apart from these difficulties in the actual practice of science, there are more substantive grounds based on philosophical reasons to question the standard assumptions of science. It concerns the way reality is described, our notion of fact and theory and relation between the empirical and the theoretical.

Classical physicists described our surrounding in a language, which neglects the relation between the observed objects. Common sense view i.e. the way layperson views the world assumes stable and unchanging things. Not only that we base our experiments upon them. But the theory of relativity and the quantum theory have shown that not only this language, this mode of perception and this manner of carrying out experiments rested on cosmological assumptions, which are not explicitly formulated. We do not also notice the assumptions and take them for granted and therefore simply speak of empirical "fact".[16] All apparently empirical facts are not only theory-impregnated as Popper thinks but thoroughly theoretical. Feyerabend raised a serious question that the dichotomy of real-unreal, objective-subjective is associated with external/objective and internal/subjective. Mind is regarded as a receptacle of everything that cannot be accommodated in the "real world". This division is also embedded in our language. "Imagining things" and "seeing something (with my own eyes)" assume that the former refers to appearance (unreal) whereas the latter refers to real objects. He also thinks that accepting some phenomena as real and rejecting others as deceptive or unreal therefore meant choosing one tradition over the others. To him arguments are rarely sufficient for getting an idea accepted (and financed to pursue) - the more important thing is that the idea must fit the ideology of the institute and must agree with the ways in which research is done there. Of course there are always institutional restrictions or elements involved in scientific research (like all other research areas) - but that does not necessarily mean it is the sole factor that determines the nature of scientific research. Feyerabend went so far as to challenge the nature of reality as offered by science apparently for the reason that there can be many possible alternatives.[17] He offered that human beings themselves might decide that they *are* really different from the reality described by the scientists. For example the quality of life can determine what is to be regarded as real and what is to be regarded as an appearance or a mere instrument of prediction.

It is possible however to view this problem from a different and much more acceptable stance in Feyerabend's sense.[18] The position can be formulated as

---

[16] P. Feyerabend, Farewell *to Reason*, pp. 290-291.

[17] Ibid., p. 262.

[18] M. Chaudhury, 'Common Sense Realism and Scientific Realism', paper presented at a seminar on 'Reality in Science and Philosophy' in February 1996, in Calcutta, organized

follows: Common-sense realism represents the view that the objects of our knowledge (mostly perceptual experience) like apples and oranges have independent existence. In other words that there is a mid-independent order of things is an assumption we take for granted in our practical life. However the presupposition of a mind-independent order of things becomes suspect to a reflective mind. How do we know that this is an apple, which does not need the knowing mind for its perceptual properties? If its existence is objective, why again there are different appearances to the same observer at different times, and different to different people? How illusion, hallucinations etc. are possible?

So a reflective approach would be one that would take into account the complex processes of various sorts that intervene between the world and the knower. 'Idea', 'sense-data', 'causal processes' etc. are some of the concepts posited to explain many things, which no direct realism such as common-sense realism alone could.

Scientific realism is the view that the objects of scientific enquiry exist and act for the most part quite independently of scientists and their activity. So defined it might be supposed that the question of whether or not natural science is 'realist' can only be settled empirically, viz. by determining whether or not scientists believe or behave as if the theoretical terms they employ possess real referents independently of their theorizing. We need therefore to make a distinction between the (relatively) unchanging real objects, which exist outside and perdure independently of the scientific process and the changing (and theoretically imbued) cognitive objects, which are produced within science as a function and result of the practice.

In short, common-sense realism insists that if there are inter-subjectively shared perceptual experiences of macro-objects like apples and oranges then there *are* apples and oranges. Scientific realism on the other hand is concerned with internal infrastructure of things, i.e. decomposing macro-objects into their structural elements, protons electrons etc. Moreover scientific realism appears to draw its high status from the purported "Omnipotence" of physics. It assumes that all meaningful questions or problems can be settled at least in principle by physical theories. This unquestionable competence of physics (or other sciences) however is contestable. There are various problems and questions that cannot be resolved by science. Creative or intellectual activity including theory-generation cannot be given a full "scientific" account. Garry Kasparov set a good example of it after beating the Super-Computer "Deep Blue" in a chess competition although he is defeated later.

Even if the micro-level analysis is corroborated by actual scientific practice it does not make the ontological status of a macro-object any less stable. In other words it becomes mandatory to make the distinction between the (relatively)

---

by PHIPC in collaboration with S.N. Bose National Centre for Basic Science, Calcutta. Later published in e-journal *Dogma*, December 1999.

unchanging real objects which exist outside and endure independently of the scientific processes and the changing cognitive objects as said in the beginning, that are products of the scientific tradition. Common sense realism speaks of macro-objects which are independent of the scientific processes, whereas scientific realism posits micro-objects which are cognitive objects within the scientific tradition and thus changes with the changes in scientific theories. I have argued elsewhere[19] that our common sense notions of space, time, causality and objects of everyday experience are not applicable either to the mega level of celestial sphere or to micro level of reality. But there is no standard common ground to judge which one is really real or more real than the other. One may recall that Galileo's findings through telescope were not instant success. At first people regarded it as "magic" compared to naked eye perception. Non-Euclidean geometry does not negate Euclidean geometry, which is still a science of terrestrial space, but only shows a way to understand celestial space of spheres. Similarly sub-atomic level reality cannot make our common-sense objects like apples and oranges (which still obeys the laws of classical physics) unreal or at least not the proper cognitive object. It is also difficult to decide why protons and electrons would be of a higher ontological status than apples and oranges. After all cognitive systems change with new *hypotheses* and new set of theoretical entities. But common-sense objects seem to be the same through different sets of theoretical entities. In brief it is difficult to dislodge common-sense belief in an external world with stable objects by any scientific theory however entrenched the latter may be.

The basic purpose of the above mentioned point is to show that it is difficult to establish that the scientific standpoint is the best or at least better standpoint than common-sense view. So even if we discount the *actual* historical processes of how scientific research is done over the years (which the sociologists of science in general and Feyerabend in particular emphasized) - scientific realism even if true cannot falsify common-sense view of the world with its relatively stable brute facts.

To come back to the main issue of scientific/expert knowledge and the alleged tyrannical elements along with its paternalistic role, one should make a distinction between (to use a cliché) ideal science and actual science. It may be a historical fact that the practice of science *has been* found with closed, partisan attitude to conform to an agreed upon standard, discounting the diatribes against its practitioners aside. But the question still remains to be answered. Can actual practice show either the goal's defect or else that there are no goals at all? If most people abuse a profession (that again is contestable), does it speak *against* the profession? Obviously the answer would be in the negative. Because norms or standards should be distinguished from the actual performances. So the standard objections that scientific and expert knowledge have been tyrannical and

[19] Ibid.

detrimental to individual freedom (information, decision, choice of method etc. that concern one) cannot adequately establish that scientific knowledge *is* tyrannical and should be either replaced by ordinary common-sense knowledge or else supervised by "laymen" or democratic institutions. Both alternatives, we shall see later, involves further problems.

However there is a more substantive problem internal to the practice of science which apparently poses serious difficulty. The problem is that (except very few naive inductivists) every sensible theory of science would regard that there must be some ideology behind every theory that decides or at least influence the kind of theory/hypothesis that would be posited. Even the choice of scientific problems, nature of initial conditions, auxiliary hypothesis etc. is at least reliant if not definable in terms of an ideology. So even granting that all scientists are pursuers of objective truth (and dismissing the charges that they are not) - the problem still remains that all methodologies are ideology-impregnated and so it is inevitable that scientists remain victims to their ideology (or dominant world-view, paradigm if you like) and can never observe from an independent stance. Indeed ideology can intrude even resulting into faulty observations.[20]

Without going into the squabble whether science is an open or a closed tradition and its practitioners rational or tyrannical - let us distinguish at least between the norm of scientific research and the actual practice. Indeed the scientific rationalists might agree that scientists are *also* human beings - of particular socio-cultural ideology "and *qua* human being" they can be "bought, swayed and deluded. But not *qua* scientists" as Noretta Koertge very aptly puts it. True scientists by definition follow the scientific method and the true spirit of that method is to make science independent of the non-rational factors like personal whims, petty interests, ideology etc. of its practitioners. The fact that ideology (or "conceit, prejudice, lie") often creeps into real -life scientific practice - is a failure of the practice not of the method or its norm. If a person commits a breach of trust, it is *his* personal defect, not the failing of the norm. If some one's action is unjust that is balance of fairness is tilted in it, this only shows that the actual act *is* wrong, not that there is no such thing as fair deal or justice. There are two standard objections: (1) content of scientific knowledge is fallible and (2) the method of science is *not* rational because these are nothing but strategies of expedience of partisan people to somehow justify their ideology – are not very strong arguments. As to the first, no philosopher of science would claim that scientific knowledge is infallible nor say that expert knowledge never failed. The second one is a more serious charge namely as science is also not a rational method it has no epistemic authority over other traditions and practices. * (Steve

---

[20] S. T. Gould, 'Morton's Ranking of Race by Cranial Capacity', *Science*, 200, 5 May 1978. The paper provides a good example of nineteenth century data on the relative sizes of African and Caucasian skulls. It is common knowledge now that 'race' is merely a cultural concept, there is no intrinsic' basis for such difference.

Fuller) We can discuss the issue with the example of Popper's methodology of falsificationism which can reasonably be called ideology-impregnated since it espouses a norm (objectivity, rationality and truth) and a goal for scientific research.

Popper's methodology of conjectures and refutations assume that the goal of scientific theory is to give a tentative solution to a problem. The problem arises not in an intellectual vacuum but in the context of extant theories, myths etc. constituting background knowledge. Hypothesis is posited by trial and error, and a tentative theory is reached when it is "severely tested" and corroborated by basic statements. On his account an observation report cannot be used to test a conjecture unless it is a "basic" statement, basic not in any ultimate sense but in an agreed upon and non-controversial sense. However basic statements also are fallible and the error can be due to the observer's prejudices or expectations. How can then this methodology claim to be a rational one? It can claim to be a rational methodology by admitting the possibility of ideological influence (even trite statement like `this is a glass' involves theoretical element) but at the same time specify some safeguards against such intrusion of prejudice or extra - rational elements into science. The recommendation that scientists deal with explanatory problem blocks or at least assigns low priority to questions like: `Are Blacks/Non-Aryan people more criminal minded /inferior in brain capacity than the white Aryans '? Explanatory problems would be rather an event or an observation report that belies our expectations based on background knowledge plus existing theories. For example if we expect a body, say a planet to behave in a certain way (circular motion in an orbit) and it *does not*, then that calls for an explanation of the problem. Tentative theories or conjectures can be offered, some or all of which may be prejudice-tainted. But requirement that scientific conjectures be highly testable, i.e. with rich empirical content, forces people make their prejudices precise enough to be refutable. This stricture on hypothesis-formation is a methodological requirement. Some examples can also be given from history of science where scientists started their research with high ideological commitment to one thesis but were eventually led through empirical investigations to change their mind.

The above case of goal/norm-oriented methodology is mentioned to show that ideology or even prejudice as such cannot make a theory irrational or its practitioners tyrannical. As a method science can have its built-in mechanism for rational evaluation. In general scientific problem arises when our expectations are violated like the case where either a theory or an observation clashes with our familiar metaphysical framework. Ideally a scientific research programme should strive to solve such cognitive problems by either severe testing of the theory in question or if the attempts persistently fail then to change the framework itself. Examples[21] can be cited where scientists started their research with a high ideological commitment to one thesis - but were eventually led through empirical investigations to change their view. Moreover even if this ideal condition does not prevail there are methodological checks such as the use of placebos, double-blind tests, instrumentation, etc. which are tactics to make the empirical basis of science as inter-subjective and acceptable as possible.[22]

The question of alleged tyranny of scientific or expert knowledge we seem to face boils down to a tyranny of logical reason over practice or scientific reason over ordinary reason. Feyerabend in this context rejects the authority of logic. Part of the value of logic is supposed to be that it provides objective standards for appraising how people think. The authority of logical rules therefore *cannot* be tyrannical, on the contrary logic or reason by definition is opposite to extra-logical norms or preferences. Logical rules/reasons are different from bias or other rules in the sense that they are *not* made by anyone to impose on others to follow. Logical rules are already there to be discovered by us to understand some situation. 'All flowers are flowers' is logically acceptable (to be true) to everybody whereas 'All flowers are beautiful' seems to be true but *not necessarily* so. That means the statement is empirically contestable. The authority of logical rules is extracted from successful practice. As Nelson Goodman[23] says principles of deductive inference are justified by their conformity with accepted deductive practice. Indeed one of the main aims of logic is the systematization of arguments in a formal system. So one necessary condition of the justification of arguments is conformity to logic, this is no draconian law - just the basic requirement of consistency of thought. Even one who rejects rationalism need the logical principle of identity ('a is a') and non-contradiction ('a is either be or not -b'). Moreover there are growing recognition of logic such as Deviant logic, Intentional logic, Fuzzy logic that take into account various shades of gray between the traditional black and white dichotomy of truth-falsity, validity-invalidity.

## *IV*

The various possible reasons mentioned in the beginning of section III for considering science as tyrannical can be analyzed below before passing on finally to the issue of expert knowledge and freedom. Although I have put the allegations

---

21 Noretta Koertge has cited three cases "randomly" chosen where scientists with certain ideological commitment have to change their mind in course of their empirical investigations. 'Ideology, Science and Free Society', *Beyond Reason*, G. Munver (ed.), p. 233-234, for these cases.

22 Placebo, double blind etc. are common methods to check accuracy of an experiment.

23 N. Goodman, *Fact, Fiction and Forecast,* pp. 63-64.

fairly separately I may bunch them together in answering because basically the allegations are empirically uncontestable. Indeed choice of problems (significant for financial crunch), priority scale and most importantly, goals of research were often contaminated by extra-scientific factors. It is also undesirable that scientists were often employed to research for some vested interests. But all these can never show that the results of these researches namely the theories *cannot* be rationally adjudged. Similarly it is undeniable that alleviation of human suffering was never the main focus of research and prestigious areas like space research and nuclear science get priority. In an article in *The Times* on the auspicious day of July 1969, when the whole world was breathlessly awaiting the first human landing to the moon - Bertrand Russell wrote an article 'Why Man Should Not Go to Moon'. There he analyzed the endless ambition and greed of men in power to conquer nature in the name of scientific research at an enormous expense, a fraction of which could have alleviated a great amount of human hunger and suffering. None can disagree with the great philosopher on this issue. Research on human problems[24] also gets low priority and therefore less fund than prestige areas like nuclear, space and computer related studies which are only indirectly related to human welfare. The charges of having male bias or being a closed community of matchmakers to a dominant theory also are not fully groundless. Ideology of an age or political regime often polluted scientific research and some results were overblown or undermined according to the dominant social/political bias.

Conceding to all these historical evidences of science and its practitioners however does not establish that science is intrinsically a social institution like any other tradition to be influenced by other social institutions such as ideology/politics or policy of the ruling power. It can at most say that actual practice of science belied its goal of finding out truth and on the contrary has been found often to be a handmaid of politics. The allegation of professionalization again is true but inevitable. It is a social phenomenon that lot of influence, lobbying, love of power etc. are motivations behind research, nevertheless the value of the research does not depend on the (bad) motive but its intrinsic worth, its contributory value to the present reservoir of knowledge and most of all on its capacity to solve some problems and predict some possible state of affairs.

Moreover empirical failure cannot be a serious ground, because no scientific theory, even observation reports claim infallible truth. The claim of expert knowledge about methodological superiority again has become a suspect. Feyerabend, the great disbeliever for example says that[25] 'my intention is not to

[24] Even in human problem areas research priority is guided by non-humanitarian norms. Infertility (which afflicts few people) gets more attention (and fund) than research on Aids, malnutrition, famine in certain countries, malaria (which is a serious menace in Asia and Africa). See the article in *Time,* June 1996.

[25] P. Feyerabend, *Against Method,* p. 32.

replace one set of general rules by another such set: my intention is, rather, to convince that *all methodologies, even the most obvious ones, have their limits'*. We will soon see this to be neither a great claim nor a claim that scientists are not aware of. The whole issue of expert knowledge and ordinary knowledge can be viewed broadly as one of reason and practice. It is true that the formal and the informal sometimes coincide, but not always. Feyerabend for example wants to show that reason is not a "disembodied" guide but a practice. And it is possible for any one practice or tradition we could come up with many methodologies or "standards of rationality". Significant periods of history of science are marked as "inductivist", "hypothetico-deductive", "methodology of research programmes", etc.

But this pluralistic picture does not show that they cannot be compared or adjudicated, as the plurality of different standards of morality does not necessarily show that all the standards are *equally* good. Indeed one can concede that science is *not* one but *many* practices over the years and over different traditions, but the practices nevertheless can be rationally evaluated. That is the main point of methodological appraisal and I shall try to formulate some common external and objective methodological standard that runs across not only different standards within scientific tradition but also outside it. But first the question of primacy of practice (tradition) over reason.

We have already seen that if logic and rationality are taken in a very narrow sense then of course many practices including scientific research are not logical. As Goodman (mentioned before) refers to principles of deductive inferences as some principles inherent in the successful practices we perpetuate, there is a tacit reliance on the basic laws of thought by all practitioners, scientific (in a narrow sense) and unscientific (in a wide sense) to understand that, we need only to consider the process of theory-construction. I am using "theory-construction" in such a way that it is applicable both to scientific tradition and non-scientific ones. When do we make a hypothesis/theory? Plainly speaking when some event or fact seems to be contrary to our expectations (born of background knowledge, extant theory and apparent observation processes). In scientific practice we *need* a (new) theory when the existing state of knowledge *canno*t explain some fact. In ordinary practice we conjecture a hypothesis when the situation looks puzzling. For example if my book disappears from the desk of my locked study, I look for some possible explanation of the event by guesses, such as (i) somebody must have got a duplicate key, (ii) there must be a secret door or window somewhere, (iii) some spirit or ghost must be behind the theft and so on. In other words (i), (ii) and (iii) are to be judged as possible ways to solve the riddle. That means theories are attempts or conjectures to solve some problems. We need to fall back upon Popper's schemata here that theories are conjectures to solve some problems. So all scientific practices start from some problems, not from a complex of observational data. His formula: Problem -> TT -> EE (criticism) -> T1. EE T2 shows basically how we proceed from a problem to a tentative theory that attempts to solve it by trial and error, which will be again replaced by a theory T2 if T2 gives a better explanation and solves more related problems than T1. Both T1 and T2 however are conjectures to be given up when they are either refuted or replaced by a better theory T3. Knowledge grows in this way through conjectures and refutations. Now there are many further problems regarding the methodological norms of 'better' theories, 'refutation', 'explanation' etc. which are controversial. But I can salvage at least one component from the morass of methodological swamp namely the yardstick of problems and attempts to solve them. If we agree that both in science and in other traditions we look for an explanation when there is a problem and that is a minimal assumption to be granted -- then it is possible to argue that we have *at least* one methodological principle to make a distinction between theory/practice which is rational and acceptable and which is not.

A theory that offers a successful solution is definitely worthy of acceptance and better than others that cannot or cannot do so effectively or completely are. But the acceptance of it does not claim however that it is logical, objective, true or infallible. It makes a very modest claim namely: given the present state of knowledge, so far this is the best, but that claim does not make any further claim either to objectivity or to truth. However the claim to accept it nonetheless is *rational*, as rational as the competent detective's acceptance of a hypothesis as the most acceptable one after judging all other possible hypotheses and the available evidences to support them. Let us examine both the points in favour of this minimal norm and also those against it.

The first objection could be one about the very status of the norm of problems. One might say that we are looking for a methodological standard, which is uncontestable, but have landed into practical success of a theory. That gives us at most a naturalistic and not a methodological criterion. Indeed the criterion of problem solving can give only a pragmatic justification, which is far from the notion of reason as a neutral arbiter among rival theories. A theory offering a successful solution can only be called a practically useful one different from the others in their practical failure, not necessarily better in other rational features. If some rule of thumb or random attempts solve a problem, that has to be regarded as a methodological rule. This is surely a serious difficulty if we regard problem solving as both necessary and sufficient for being a methodological standard. But a weaker claim can be made that it is only *necessary* and not sufficient that a scientific theory has to solve the problem in addition to other rational feature. Whatever way the problem is solved however cannot be called scientific theory. Then random efforts, rule of thumb, accidental success etc. in solving a problem cannot necessarily be called an integral part of a theory. In other words an acceptable theory must solve a problem at least equally and ideally better than any others must but problem solving by any *ad hoc* means cannot be a good methodological criterion. We should however clarify another point here to avoid any simplistic account of problem solving as found in the pragmatist view. As I said before we suggest a theory when there is a *problem*. So in a sense 'science begins with problems' as Popper[26] says. We attempt to guess a solution to our problem and 'we attempt to *criticize* our usually somewhat feeble solutions'. 'Sometimes a guess or a conjecture may withstand our criticism and our experimental tests for quite some time. But as a rule we find that our conjectures *can* be refuted or that they do not solve our problem, or that they solve it only in part'.[27]

This interpretation of a problem-situation is crucial for understanding the methodological criterion we are upholding. It entails that even the best solutions arrived at by the most severe tests and the most sincere and brilliant mind may soon give rise to new difficulties and new *problems*. As said before, expectations (even new-borns have some) if thwarted-give rise to problems -- both theoretical and practical, and we can try out a conjecture to solve it with the help of tests and experiments. Popper rightly thinks that experimental tests are also part of the process of 'working on a problem', i.e., getting acquainted with the problem and its various aspects and by this process also 'improving ones chances of finding ... a solution'. To understand a problem thus implies a possibility of finding a satisfactory solution – but such understanding by no means entails that a solution in fact is found or surely to be found. The significance of such an analysis lies in its recognition of complexity of a cognitive situation, which cannot be reduced to its empirical success. Cognitive processes are much more complex than a successful end-result. Recognition of a problem-situation indeed involves theoretical elements, but attempts to find out a solution involve enriching cognitive processes even if such attempts fail to achieve success. Such analysis is neither simplistic, nor naturalistic because it makes identification of a problem itself a methodological standard along with its attempted solution. It is different from a pragmatic approach because empirical success is *not* emphasized as the only external norm to evaluate a theory.

A more serious objection however can be raised concerning the very nature of the concept of what is a problem. One may argue[28] that it is begging the question because what counts as a problem depends on the theory. There are no brute facts as problems, the latter occurs as mentioned before, when one's expectations (fed by background knowledge, theoretical framework etc.) are not fulfilled. But those expectations are bred within the domain of rationality of the framework. Therefore problems also cannot arise in an intellectual vacuum. There is one way to get out of this fix: to distinguish different problem areas and then show that

---

[26] Karl Popper, 'Science Problems, Aims, Responsibilities, sec. VIII, IX, X, reprinted in, *The Myth of the Framework*, ed. M. A. Notturno.

[27] Ibid. p. 96, emphasis added.

[28] J. Kekes, 'Reason and Practice', in *Beyond Reason,* ed. Gunver, pp. 159-78. Kekes also regards problem solving as a good practice over reason but in a pragmatic way, not in Popper's methodological sense.

within each of these areas human beings encounter the same problems regardless of what theories they hold. This is a pragmatic move based on an anthropological argument. J. Kekes holds such view, and 'problem of life' is a key concept in such an account according to which 'rational practices must directly or indirectly contribute to solving problems of life. The problems of life occur, according to this view because human nature has some unchanging element. This invariant component provides the standard with reference to which practices can be rationally judged. But notice that the concept of rationality here is intra-theoretic and context-dependent. Of course it is opposed to relativism or anarchism of epistemic sort for a theory that succeeds cannot be at par with another practice that fails to solve a problem. But this judgement is not imposed by any epistemic authority. Then according to this view *all* traditions which can solve a problem are at par and looking for *better* or more comprehensive theory will not be the goal of scientific research. Progress of knowledge also would be a redundant metaphor.

The other way that I suggest is to make a distinction between problems that are practical and those which are theoretical/cognitive problems. The latter kind of problems of course are theory-generated, therefore solutions of them should be in terms of the theoretical framework they belong to. Practical problems also involve theoretical elements but only indirectly. Even innocuous and seemingly uncontroversial basic statements like 'this is water' have a theoretical content. But practical problems differ from theoretical problems in one important aspect. The aim of practical problems lies *entirely* in its solution, it does not necessarily aim also to have a theoretical justification for it, the theoretical problems on the other hand not only call for an theoretical explanation but also for tying up processes of logical/rational aspects of other principles of the framework for an adequate solution. The theoretical problem with the apparent anomaly in the orbit of the planet Uranus when tied up with the theoretical framework of Newtonian theory of gravitation, circular motion of planetary bodies, the computation of orbits of Uranus and other bodies eventually led to the solution namely the discovery of Neptune from the rational explanation 'there must be some hitherto unknown body in the orbital path of Uranus'. Practical problems do not always demand a rational explanation, although there can be some, but often a solution itself is enough to solve what Kikes calls "problem of life".

The advantages of problem solving, as a methodological arbiter is that other methodological strategies are dependent on the tradition and also on the dominant theory. They imbibe even the vocabulary and presuppositions of the tradition. In other words all methodological standards are internal to the standard (the sociologists of science are right about this). But problem solving is *external* to traditions. It is a standard of acceptance if not justification external to traditions and therefore one which can consequently rationally evaluate traditions and standards inherent in the traditions. As I said before cultural or ethical relativism only espouse that there are many traditions, but not also that these traditions cannot be adjudicated. All other standards, one can say are human inventions and thus internal to the traditions but problem solving is *not* a human invention in any sense. Popper also regards problem solving as the prime urge from which first myths and then science has been generated. This analysis will also encompass the Feyerabendian tirade against abstract logic and take into account his substantive thesis that methods (of science or any other tradition) are devised in order to cope with concrete complex problems, but does not share the other part of his diatribe that scientific method is nothing but shrewd tactics full of conceit and propaganda of opportunistic people for politically winning intellectual battles in any way including unfair means.

The main substantial objection Feyerabend raised against science is that it cannot be given preferred treatment by society on the ground that it and it alone follows the dictates of reason. Because the *first* problem is that science does not adhere to the standards so far proposed by philosophers of science. The *second* is that even if we found that science accords with reason, reason is dependent upon practice and thus not a suitable arbiter of tradition. I have argued above that problem solving as a methodological standard (nobody can deny the primacy of problem in any tradition including science) can serve not only as a good arbiter but also as óne to favour science for good reasons.

## *Concluding Remarks*

The anxiety that scientific/expert knowledge has reached such a height in specialization that individual freedom and ordinary reason are at stake – is a genuine concern. In different areas of knowledge and life science and its practitioners enjoy a cognitively high status at the cost of other traditions. It is felt by many that this unquestionable authority of scientific knowledge over other practices can be challenged and other practices placed in the right context. Some thinkers also hold that ordinary reason (if not better) should be treated with the same kind of respect as scientific reason.

The high status conferred on science through many historical events resulted into its role as a neutral arbiter of true and the rational. And this sacrosanct authority empowered science and the practitioners of science to the extent of tyranny. This concern is genuine and it is to a great extent true that scientific knowledge has often played a tyrannical role both cognitively (disregarding other alternative traditions) and empirically. For these reasons scientific knowledge has become a suspect to many people.

Granting all these to the critics, we must at least distinguish between what science aims for and what actually it has achieved. Even granting that the tremendous power of scientific achievements have been exploited for political reasons and not so much for alleviation of human suffering, one may say that all these could only establish gross misuse of science and *not* that the latter is an irrational tyrannical institution. I argued that basic law of logic and truth is the general aim of any scientific enterprise. "Scientific" stands for a stance that is based on logic and reason compared to any other standpoint where other considerations may prevail, although it is arbitrary to mark modern western science only as science and all other traditions as "unscientific" or "pre-scientific".

We should emphasize in this connection that the alleged tyrannical element lies not in the intrinsic nature of scientific knowledge but rather in its misuse or abuse by political power. However the general resentment is justifiable that most of the time people do not have enough information about scientific achievements and decisions. More transparency about scientific activities and other details (such as expenses, time and the risks involved) is desirable in a free society. The single example of the jury system (Feyerabend's favourite example) can show that the intuition and reasoning capacity of ordinary people should not be underestimated. In complicated legal processes like jury system works quite satisfactorily.

But this argument is not applicable to scientific knowledge and decisions in the same way. For the latter involves lot of background knowledge with technical details. The actual research processes are tedious, repetitive perseverant routine activities of trial and error. It involves long equations, complex calculations etc., which are not only uninteresting but also unintelligible to one outside the specialized circuit let alone the ordinary people. Only the end result if successful in any mundane way can interest outsiders. A space research scientist in the mix-sixties when asked about his research project answered with a smile that 'without the tedious details; if it is successful then people from every part of India can watch the same television programme'. Surely we realize today that his project had been successful!

Another way we can understand the situation is by distinguishing between scientism and scientific knowledge, expertise and expert knowledge. Scientism is "aping of science" as Hayek calls it – that is unnecessary stretching of science which is a caricature of science. Expertise similarly is a particular skill in a particular area whereas expert knowledge ideally is a specialized knowledge that can interconnect (and sometimes solve) a problem with many other interrelated areas. It is desirable indeed that expert knowledge should be explained in terms that are easily understandable by any one. Otherwise a decision would appear to be unilateral and tyrannical due to lack of transparency. Medical or ecological cases particularly are in need of such dissemination of scientific knowledge.

One last but quite significant point is there to be considered which works from within the scientific tradition. As I mentioned before, the objectivity claim of science is rejected from within by the indeterminacy principle. No tyrannical institution will specify the conditions under which it has to be rejected. But as argued before, there are methodological devices such as placebo, double blind tests, instrumentation etc. which act as safety guards from within the scientific tradition. It is similar to the blind referring system for publishing to check editorial tyranny. This is not to say that the actual practice *is never* tyrannical, but only to argue that an institution as such cannot be termed tyrannical if it also includes methods to inhibit or at least minimize possibility of tyrannical treatment.

Finally the problems related to domineering role of science and technology cannot be overestimated. Feyerabend's fiery diatribe has many substantive points. It is time one should be aware that western tradition is engulfing all other traditions. Even concepts and language is entrenched by it. Why for example western (mainly Anglo-American) culture is 'mainstream' and other cultures are 'alternative cultures? Surely eastern civilization has at least some claim to be temporally prior. The division into 'First world' – "Third World" countries, "developing" (a patronizing change from "underdeveloped" and "developed" also is arbitrary and questionable. Feyerabend and many others are also right that western science has been exploited more for accumulation of power than any human cause. Two areas can be particularly identified where individual rights are not respected. These two are medical profession and education where patients and students are exploited because transparency and accountability is lacking. It is not clear how to determine the success of a physical/mental or intellectual treatment.

Granting all these to one who challenges this paternalistic picture, it is not clear that the picture can be reversed without violating the basic tenets of freedom. I have already hinted in sec. I that Mannheim's 'militant democracy' to monitor all scientific decisions cannot be a viable democratic norm. Similarly Feyerabend's slogan, 'laymen can and must supervise Science[29] is not tenable either. He thinks that 'the errors of specialists can be discovered by ordinary people' only if they are prepared to do some hard work. There should be 'duly elected committees' to 'examine the safety of nuclear reactor's or whether 'scientific medicine deserves the unique position of theoretical authority'. He thinks 'scientists, educators, physicians' must be supervised not only when engaged in *public* jobs; but must also be 'watched most carefully' when called upon to solve the *problems* of an individual or a family in somewhat similar way as we watch plumbers and carpenters after comparing estimates from different firms.

It is undeniable that there is a patronizing element in the way in which specialists treat layperson. But supervision or 'watching carefully' cannot solve the difficulty. Apart from the practical problem of understanding the technicalities in which plumbers and carpenters also excel in deluding laymen – there is some internal flaw in Feyerabend's argument. When he says "duly elected committee of laymen" or people "prepared to do some hard work" are qualified enough to do the "supervision" of specialists, there is a tacit distinction made between other laymen and these "duly qualified ones" which amounts to saying that the latter

[29] P. Feyerabend, *Science in a Free Society*, p. 96.

category is specialized to supervise. So we come back to the notion of specialist again only in the sense of a kind of super specialist to watch ordinary specialists. Secondly if ideology creeps into scientist's theory because they are human – these "supervisors" are equally vulnerable to similar prejudices, more so because they are aware of their "policing" role. Scientists at least face many checks internal to the system, but layman-supervisor has only political ideology and therefore more chance of being prejudiced.

Besides these difficulties there is another more serious problem I see in such supervision to protect individuals from tyranny. My objection is that it is self-defeating. The practical problems aside, such project defeats the very cause it aims for. The basic point here is: expert scientific knowledge has in actual practice been often tyrannical. Now supervision is suggested as a way to watch if there is any infringement of individual freedom. But scientists are also *free* individuals and their enterprise should be a *free* enterprise at least in a free society. No cognitive projects especially scientific research can be freely pursued if there is constant supervision or at least "careful watching". It will not only be a loss to progress of knowledge (no one can work or achieve successfully if there is someone breathing over her shoulder) but also infringement of individual freedom. It is a grave infringement since the area is particularly important for its creative and cognitive value.

Therefore supervision or monitoring cannot be viable methods to combat tyranny if there is any. It should be dealt with from within the tradition by methodological devices and transparency. Any such attempts from outside the tradition will only lead to more serious troubles including violation of freedom thus defeating the very aim these are designed for. It is clear that the image of scientific research as a course of unilinear progressive nature is oversimplistic. Growth of scientific knowledge is neither steady nor monolithic. It can grow in diverse directions, stall sometimes and even include periodic regressive phases. Human interests and social factors also can influence its direction and priorities. It would be a myopic methodology if other traditions were ignored and human factors overlooked. Instead of either having an attitude of reverence and complete trust on scientific experts or on the other hand undermining their positive achievements, a comprehensive appraisal of the issue of science and other traditions should be undertaken.

First: we should examine if attention and focus of scientific research is proportionate and reasonable from humanistic standpoint. Second: the more controversial area of research is that which concerns human problems directly (medical treatment, nutrition, birth control, genetic engineering, etc.) rather than areas remote from human concerns such as space-research, nuclear science, geological and astronomical sciences. Finally the battle between scientific and other traditions or the expert-lay people controversies do not lead us anywhere. Instead of scientific expertise vs. other culture we can say scientific expertise *and* other cultures in a cooperative sense. That kind of concept is not absolutely unknown - it is already embedded in our notion of interdisciplinary studies and enterprises.

### *(i) Focus of scientific research*

The most prevalent reason for resentment about scientific research tradition is that it allows disproportionate amount of attention and research funding to areas less vital or major problem area for most. Some problems that affect comparatively few people (e.g. infertility) receive more attention than problems that seem to be global in comparison. AIDS, famine in some countries, drought or flood conditions etc. 'continue to simmer on the scientific back burner'. Indeed some attention and money have been spared on these problems to certain extent but that did not help much. So a totally new approach is needed that requires a breakthrough from the traditional disciplinary barriers. It appears that enough attention (and therefore funding and time) is not given to many global problems concerning human welfare comparable to the enormous amount of research time and money spent on other so-called prestige areas such as space research, super conductor, sophisticated armaments and defence equipments. One single example can indicate how a very major human affliction is neglected with sheer indifference by scientific community. The case I refer is the case of malaria. Half a billion people catch malaria every year. There are other mosquito-borne diseases also (dengue, filaria, etc.) which claim thousands of lives every year. Quinine, the traditional medicine to cure malaria remains still to be expensive to harvest, synthesize and fails to prevent relapses. Attempts by the World Health Organization to eradicate the disease by killing the courier failed when anopheles mosquito became resistant to pesticides such as DDT. Almost half of the world's population is still vulnerable to this disease and desperately in need of a cheap and effective vaccine. But modern drug firms seem to ignore the need and lack motivation for related research efforts.

Western companies manufacture number of preventive medicine for rich tourists to malarial climes. Chloroquine for example provides (in combination with some other drugs) almost 70% protection against the malarial parasite plasmodium. In higher dose it is also effective in curing patients who have malaria. But these drugs are far too expensive for those most at risk, 90% of malarial victims being sub-Saharan Africans and moreover plasmodium is fast becoming resistant to them. Mefloquine is a stronger prophylactic than Chloroquine and is effective in 90% cases but it is even more expensive and has side effects (a British Medical Journal study reports that one in 140 users suffers incapacitating dizziness, panic attacks, hallucinations or other neuro-psychaiatric problems.

Given the enormity of the malaria menace (half a billion catching it of whom more than a million dying every year) it seems that malaria research should attract more attention. But the world spent less to research on the killer protozoa than even on asthma and AIDS. It seems governmental research agencies show more effort than private pharmaceutical companies. The real reason behind the apathy is commercial. For most of the victims are too poor to pay being from Asian and African countries. And if the tourist market is the target the risks are high because in that case the use is preventive only for well-to-do customers from developed countries. In case of any failure or adverse effects they can bring litigation against the companies. But this neglect in the focus of research in Malaria is shortsighted because with the steady changes in world climate malaria can also come back to western countries from which the killer disease left only a century ago.

### *(2) More controversies in human areas than in non-human areas*

It has been already pointed out that problem areas that directly or indirectly affect human being are not the priority areas of research and they are also subject to controversy. Space research, geological or astronomical sciences involve programmes whose main target is more knowledge about these areas, not solving some human problems but then the goal of science is to set the limit of what *cannot be known* and *not what should be done for human welfare.* The latter may be a secondary consequence but improving human conditions is not the principal aim of scientific research programmes. There may be some minor points of difference among rival paradigms in those sciences but not so much debates and disagreements as found in areas that concern human welfare to some extent. The reason behind the disagreements is that between the two goals of science, namely knowledge of nature and solution of human problems -- there is usually no debate about the first as such. But when the second is also taken into account there are number of normative questions that can be raised. Should we search for knowledge's sake only or all scientific research should aim at human good? Is it possible to set some moral goals for scientific pursuits or scientists should be given complete freedom to pursue knowledge? And so on. The notion of science policy therefore becomes very significant -- but again if science policy is decided by some super, extra-scientific norm, there is always a risk of either paternalism or guidance by some ideology. Most critiques of modern science point towards cases where success is not achieved or not up to their political goal. Both paternalistic and ideological goals are threats to scientific freedom. So when human problems and their minimization (if not solution) are concerned it is difficult to decide whether scientists should work in absolute freedom in search of "truth" especially because of time and fund constraints. Even a naive layman is justified in thinking why can't scientific knowledge control devastating floods, epidemics and other calamities while enormous amount of money and human resources are being spent on knowing the nature of celestial bodies, paleolithic age or intricacies of micro bodies and so on?

### *(3) Scientific expertise and non-scientific traditions*

The debate whether science is the most prestigious (if not the only) cognitive programme or merely *one* of many traditions has become a bitter issue and a defunct cliché. Indeed science as a research tradition has belied many of our hopes and aspirations. The general frustration is so enormous that now people are gradually turning to other traditions for their comfort. Instead of only condemning this phenomenon as regressive and superstitious, one should make an in-depth analysis of why anti-scientific or pre-scientific traditions are slowly gaining ground in many countries.

Prayer, Faith healing, Reiki and Pranic (both of the last two are Asian mode of healing by touch systems of cure of many ailments) are steadily getting popular even in western countries. The most astonishing turn is that some scientists are beginning to look seriously at just what benefits patients (say after a open-heart surgery) may derive from extra-scientific practices like prayer or touch and so on.[30] It is held by many that as the 20th century draws to an end, and the new millennium starts 'there is growing disenchantment with one of its greatest achievements of modern science namely its high-tech medicine. Western medicine is at its best in a crisis-battling acute infection, repairing the wounds of war, replacing a broken-down kidney or heart'.[31] But increasingly what ails people of modern societies are chronic illnesses such as high blood pressure, body -ache, arthritis, cardiovascular disease, depression, even cancer and AIDS all of which are stress and life-style related diseases. These are not merely bodily problems; therefore, other traditional practices that take the mind also into account can give relief and solace to a great extent in these cases. The effect of religion, prayer and touch etc. on healing is of course debatable except the psychological element. The traditional medical systems (Ayurveda, Unani, Homeopathy etc.) are however *not* based on spirituality and therefore are of testable nature. More research on these practices can be useful in the sense that their claim to knowledge can only be legitimized if the dominant scientific tradition sanctions their methods. One way to such recognition is through their inclusion in the mainstream educational curriculum.

---

[30] 'Healing', Claudia Wallis, *Time,* June 24 1996 gives a well-documented account of how in America not only people are turning to extra-scientific modes of healing but the scientists also are researching seriously for the effect of religion and faith related practices on patients. 'Twenty years ago, no self-respecting American doctor would have dared to propose a double-blind, controlled study of something as intangible as prayer. Western medicine has spent the past 100 years trying to rid itself of remnants of mysticism ... Today ... a shift among doctors in the U.S. toward the view that there may be more to health than blood-cell counts and EKGs and more to healing than pills and scalpels'.

[31] Ibid. p. 39.

With such idea in mind Ashis Nandy[32] says that universities in India 'have to develop a new openness to and protectiveness towards non-modern, local systems of knowledge. These systems are the ones which need protection now, not the modern ones.' He envisages the role of future universities not serving 'only as repositories of knowledge valued in the famous western centres of scholarship but they have to 'inculcate a robust suspicion towards the victorious systems of knowledge, and stress the recovery and transmission of knowledge that has been cornered or defeated.' Nandy made three propositions: (1) Recovery of indigenous knowledge is not easy because modern (scientific) knowledge is available in a "ready made capsuled form". The damage caused by mindless consumerism can be repaired by the university. (2) The dialogue between the tradition and the modern has broken down. The university can mediate between the dialogue of 'vernacular knowledge acquired over the centuries and the more elegantly packaged, laboratory-based knowledge'. (3) The job of a centre for higher learning is not to produce a perfect human being but rather for making possible a range of options for people and society to freely choose from.

No one can agree more with Nandy on these points particularly the third proposition. Indeed the aim of higher learning is to create an ambience of freedom and rationality. No good will come out of any kind of social engineering. His vision of future university as a mediator between tradition and modern forms of knowledge sounds reasonable particularly if we consider how much area of our modern life is taken out of traditional participatory style and given to experts. In traditional India most of the areas (even medical treatment) were shared by people of the community with some degree of specialization. Now experts reign in every areas, even bridal make-up and wedding management, are at the hands of experts. Therefore staying outside the modern system of knowledge would only give tradition the status of an "alternative system". To be effective, it is prudent to be part of the mainstream modern institution such as university, which is regarded as the mainstay of expert knowledge of different sort.

To conclude it is not clear that scientific expertise is not to be found in the specialized skills of its practitioner or that specialization and its associated expertise is 'simply a strategy for being a player in the epistemic marketplace' as Steve Fuller says.[33] Fuller's main contention is that rhetorical strategies used by scientists are responsible for people's trust in their expertise. The recurrent theme of this work is that the optimal way to democratize science is to detach it from any specialized status and practice. This theme is not very different from the Feyerabendian position. Democratic practice also is the common key notion.

---

[32] Ashis Nandy, 'The Future University', *Seminar*, 425, January 1995, p. 96.

[33] Kuhn, Feyerabend and most sociologists of knowledge are of such opinion. Recently Steve Fuller suggested that rhetorical strategies have effectively removed scientific decision-making from public scrutiny. See his *Philosophy, Rhetoric and the End of Knowledge,* Chapt. 2 and 3.

Knowledge production should be encouraged with maximal public involvement, open to all people, irrespective of their expertise and often despite their lack of scientific training. It is true that specialists take on a narrow view of problems and do not relate it to larger projects. So the question still remains that competence of scientific expertise is not uncontestable at least in the public sphere of global problems. But it is not clear that scientific expertise has nothing to offer in the solution of these problems. Specialized knowledge has an important role to play (say in curbing an epidemic like SARS which is raging the world at the moment or AIDS as an immunological problem) here as well. But what is lacking in this approach is the lack of coordinated effort from other disciplines as well to have a comprehensive approach to the problem. Universities so far breed only experts without this vision. In Nandy's view perhaps this lack of coordination is indicated. Thus the best way to supplement expert knowledge is *not* democratization of science by lay people (which involves further problems as discussed before) but interdisciplinary participation that helps realignment of scientific talent and other expert knowledge to a common concern. Indeed sociologists, historians, geographers, anthropologists and most importantly philosophers can supplement expert knowledge for a better understanding of problems in a global perspective. This idea is already implicit in the concept of science policy, which however needs to be redefined and refurbished from time to time.

However to critique science because of adverse effects of many technological development is wrong because scientific research is independent [34] of the turn technology takes. The different kinds of critics of modern science including the post modernist and feminist scientists jumped for Feyerabendian kind of "no-method" view because they have identified science and technology as one and the same. We should separate these two issues because not only technology is revisable but also scientific theories also are fallible and never claim certainty. But the course of development for science should never be stopped or reversed for it will then be the death of individual freedom and end of an open society.

---

[34] David Miller, 'Is Scientific knowledge an Inexhaustible Economic Resource?' *Critical Rationalism* , Section 3. Chicago, 1994.

## *CHAPTER IV*

# FREEDOM, DETERMINISM, INDETERMINISM

Determinism and indeterminism are philosophical theories that are better understood than they are formulated. The former is at least meant by different positions such as metaphysical and scientific determinism. Metaphysical determinism views that all events in this world are predetermined or unchangeable. It does not of course claim that the events are known or 'predictable by scientific means'. 'But it asserts that the future is as little changeable as the past. Everybody knows what we mean when we say that the past cannot be changed. It is in precisely the same sense that the future cannot be changed, according to metaphysical determinism'. This formulation is due to Karl Popper;[1] it also indicates that in this sense metaphysical determinism is *not testable*. Popper also says quite rightly that both metaphysical determinism and metaphysical indeterminism are untestable. The former is untestable because even if the world continues to appear as surprising showing no signs of regularity or predetermination it can nevertheless be determined. Metaphysical indeterminism is untestable because however much the worldly events appear to be regular recurrences, which are not sufficient to establish that no undetermined event can exist. It is evident that untestability or lack of empirical content makes the theory logically weak.

Popper has taken a strategy to argue against determinism although he does not allege that it is "meaningless". The strategy is like this: metaphysical determinism is a weak (due to its weak empirical content) doctrine, therefore it maybe entailed by a strong doctrine. Scientific determinism is such a theory. Thus the arguments that support scientific determinism are the strongest in favour of metaphysical determinism. If these arguments are defeated, that will be the end of scientific determinism and along with it the collapse of metaphysical determinism will be inevitable. Popper has advanced some arguments against both commonsense and philosophical arguments in favour of scientific determinism. He also offers some direct arguments against the acceptance of metaphysical determinism and some based on classical physics. But before examining these arguments let us clarify the two positions; metaphysical determinism as we see is the thesis that *all events are, at least in principle causally predetermined* while scientific determinism is the view that all events are scientifically predictable. It is clear that metaphysical determinism does not have any claim about predictability. However both the theses are opposed to the doctrine of indeterminism at least in principle and thus are also inconsistent with

---

[1] Karl Popper, *The Open Universe*, p. 8.

the concept of human freedom. For the latter concept again living in an indeterministic world is not enough, as we will see later. But first about determinism in social and physical affairs.

## *Methodological Aim and Determinism*

Looking for an adequate method of enquiry which ideally can resolve all difficulties of an area of enquiry is an age-old task. That being found too naive and impossible an expectation, methodology now stands for a modest claim to establish certain rules and requirements which are necessary to prohibit some wrong moves. But it does not however pretend to guarantee sure success. These requirements are mainly of two different kinds: formal and material. The former basically includes logical principles and some stipulated norms, while the material requirement of any methodology consists of the contents of the premises of an explanatory theory in a particular field, which a theory has to satisfy. These requirements may be called "regulative principles".[2] The metaphysical theory of mechanism can be a regulative principle, which governed thoughts in physical sciences for a long period until the wave of field theory changed the scenario.[3] According to the mechanistic view the ultimate constituents of the physical world are unanalysable particles which obey mechanical laws. As these particles are simple ultimates, their existence cannot be explained whereas every complex physical things or events are the results of particular configuration of particles and can be explained in terms of the laws determining their behaviour in conjunction with a description of their relative positions, masses, momenta, etc. In other words the explanations of all large-scale phenomena can be done ultimately in terms of statements about properties and relations of particles.

Mechanical theories are indeed confirmed with great success, but nevertheless they are untestable. That means no experiments can overthrow or refute a mechanical theory. An example (due to Watkins) is electro-magnetic phenomena, which 'seem refractory to this mechanistic sort of explanation. This refractoriness can always (and perhaps rightly) be attributed to our inability to find a successful mechanical model rather than to an error in our metaphysical intuition about the ultimate constitution of the physical world'.[4] But mechanism, as Watkins rightly observes, although non-empirical in nature, is a strong theory to be incompatible in principle with various conceivable physical theories. Thus it acts as a 'regulative' principle - though metaphysical it guides research at least

[2] J.W.N. Watkins, 'Methodological Individualism and Social Tendencies', *Readings in the Philosophy of the Social Sciences*, ed. May Brodbeck, pp. 269-280.
[3] Ibid., p. 270.
[4] Ibid.

by directing which way one *should not go*. This will narrow down the focus of research from wild guesses.

In a sense the mechanistic theory about the physical universe is analogous to the principle of methodological individualism in social science. According to this principle every complex social situation, institution, or event is the result of 'a particular configuration of individuals, their dispositions, situations, beliefs and physical resources and environment.' There may be incomplete explanations of large-scale phenomena in terms of other large-scale phenomena, for example inflation can be explained halfway by referring to full employment. But we cannot arrive at rock-bottom explanations of such large-scale social phenomena until we have inferred an account of them from statements about the beliefs, dispositions etc. and interrelations of individuals. This is the typical Popperian position and it is clear how close it is to Mill's individualism which Popper labels as "psychologism".[5] The view[6] is contrasted with sociological holism, according to which social systems constitute "wholes" in the sense that their large-scale behaviour is governed by macro-laws which are essentially sociological at least in the sense that it cannot be explained as mere regularities or tendencies resulting from the behaviour of interacting individuals. On the other hand the sociological holism holds that the behaviour of individuals should be explained at least partly in terms of such laws. In short if methodological individualism regards individuals as the guiding agents in history, holism is the view according to which some superhuman factors are the moving force. The alleged long-term cyclical wave in economic life, for example is supposed to be self-guiding, inevitable principle that explains large-scale phenomena such as wars, revolutions, mass emigration including psychological factors such as scientific and technological inventiveness. The claim of holism goes further than explaining social phenomena to prediction of future events as the sociological laws postulated by the holists claim also to be the laws of social developments, i.e., law governing the dynamics of a society. A version of it is the idea that a society is driven along a pre-determined route by historical laws, which are irresistible but discernible, by sociological analysis.

The holist historicist position is a typical determinist position in social science methodology. The central assumption of the individualist position against it is that *no social tendency exists* which could not be altered if the individuals concerned both wanted to alter it and possessed the appropriate information. Information is significant because due to ignorance they might fail

[5] Watkins maintains such a view but it is not clear whether Popper would have endorsed the analogy.

[6] Methodological individualism has been defended principally by F. A. Hayek in *Individualism and Economic Order* and *The Counter Revolution of Science* and by K. Popper in his *The Open Society and its Enemies* and *The Poverty of Historicism*. J.W.N. Watkins also defends it in principle.

to alter it or on the contrary even aggravate the existing situation. In other words, no social tendency is imposed on human beings either from above or from below. Social tendencies are the product (often unintended) of human characteristics and situations, ignorance as well as knowledge, ambition and lethargy. Indeed there are some tendencies which are sometimes controlled by some compulsive physical factors. For example in a dominantly agricultural society where very little technological development has taken place, there would be a general tendency to have more male children (who can be of more help in the field) than female children.

As mentioned before if the general aim of methodology is to find out same rules of enquiry which will enable one (if not to find out the true/valid theory) at least to focus on *some* move by elimination of various wrong moves. In that case the philosophical assumptions of determinism/indeterminism become very significant, particularly in connection with the methodology of social sciences. We will see that determinism cannot hold either in the physical universe or in society. We can examine these issues particularly with reference to the fiercest critic of determinism and staunch defender of human freedom - Karl Popper.

## *Society, Trends and Historicism*

Popper's polemic against the so-called doctrine of historicism is famous and at the same time raised many controversies.[7] I am not going to address the questions whether he was fair to Plato and Hegel, or whether his account of Marxist theory of history is correct or not. I shall concentrate instead on the central question whether prediction in social science is possible on the assumption of some deterministic law or tendencies. The basic claim of scientific determinism is that future events are predictable because they are causally related to some prior states. If scientific determinism (a stronger position than metaphysical determinism) can be shown to be false, then social determinism also cannot hold. And historicism is a typical determinist position.

To understand Popper's tirade against historicism, one should remember that he tried to refute the principle of the unity of method and to lay down special restrictions on research in the social sciences.[8] He characterizes historicism as: 'An approach to the social sciences which assumes that *historical prediction* is their principal aim, and which assumes that this aim is attainable by discovering the "rhythm" or the "patterns", the "laws" or the "trends" that underlie the evolution of history'.[9] Another crucial aspect of historicism allegedly is that in making statements about the future it employs "unconditional historical

---

[7] K. Popper, *Poverty of Historicism* contains arguments against historicism.

[8] Ibid., and also *Open Society and its Enemies,* vol. II.

[9] K. Popper, *Poverty of Historicism,* p. 3.

prophecies" rather than "scientific predictions" of the kind found in physical sciences.

The main arguments against historicism are as follows: one, 'universal laws make assertions concerning some unvarying order ... and any law formulated in any way 'must be *tested* by new instances before it can be taken seriously by science. But we cannot hope to test a universal hypothesis nor to find a natural law acceptable to science if we are for ever confined to the observations of one unique process'.[10] The basic point here is that *the evolution of life on earth or of human_society is a unique historical* process. The historicist law is, Popper presumes, of the form that ranges over complete evolutionary (historical) sequences. But as the universe of discourse contains only one element - the actual historical event - historicist law must be untestable and therefore "unacceptable to science". To strengthen his argument Popper assumes two possible objections to his point apparently with the hope that these objections can be adequately met. But it can be shown that if these objections sustained then Popper's claim that those historicist doctrines are *not* scientific, would not hold. Peter Urbach[11] argues that Popper did not succeed in replying to these objections. The objections as Popper presented are as follows:

(a) One might deny that the evolutionary process is unique and say that in many different societies the same evolutionary process can be found . In discussing this objection Popper admits that sometimes 'history may repeat itself in certain respects'. But he maintains that 'all these instances of repetition involve circumstances which are vastly dissimilar, and which may exert an important influence upon further developments'. Although he admits that there are some parallel between 'certain types of historical events, such as the rise of tyrannies in ancient Greece and in modern times' - but as these events involve "vastly dissimilar" circumstances - he concludes that 'we have therefore no *valid* reason to expect of any apparent repetition of a historical development that it will *continue* to run parallel to its prototype'.[12] As Urbach points out, Popper seems to be saying that we cannot validly infer the historicist theory from observations of particular historical developments. But why? The ground seems to be that historicism is not *proved* from fact and Urbach finds it contrary to Popper's own methodology according to which theories should transcend the given facts and still be testable. Therefore, Urbach thinks that Popper's reply to the objection that historicist theory is based on parallels is ineffectual. But Urbach's criticism misses one vital point - namely Popper rejects historicism *not* because it is not proved from facts. On the contrary he admits that it is (like many other metaphysical theories) seemingly confirmed by facts. But the very next sentence

---

[10] Ibid. pp. 108-109.

[11] Peter Urbach, 'Is Any of Popper's Arguments against Historicism Valid'? *British Journal of Philosophy of Science*, 29, 1978.

[12] K. Popper, *Poverty of Historicism*, Harper, 1957, p. 111 also cited in Peter Urbach.

is crucial. It reads '... facts which if examined more closely, turn out to be selected in the light of the very theories they are supposed to test'.[13]Therefore the confirmation is trivial as it presupposes the theory.

(b) After granting that the evolutionary process is unique - the historicist may nevertheless claim that we may discern 'a trend or tendency or direction and that we may formulate a hypothesis which states this trend, and test this hypothesis by future experience'.

This objection imagines a situation in which a trend has been observed in some society over a period of say increasing political instability and on that basis the historicist may generalize and predict that this trend will continue indefinitely. Here Popper objects to this generalization on the basis of a distinction between a trend and a law. He adds that trends are *existential*, not universal. A universal law, on the other hand, does not assert existence. On the contrary, "it asserts the impossibility of something or other".[14] Popper's main contention is that a statement asserting the existence of a trend at a particular time and place would be a singular historical statement, not a universal law. On this distinction we may base scientific prediction on laws but cannot make it merely on the basis of an existing trend. A trend may change or stop within a short period or even abruptly due to some countervailing condition. Moreover no succession of phenomena proceeds according to any single law of nature, such as that of gravity, or even according to a single set of laws. We have to consider many other laws and boundary (initial) conditions. If an apple falls, gravity would not explain this apparently simple phenomenon - we need to consider laws about wind pressure, the actual jerking movements of the branch, the tension in the apple's stalk etc. Besides the examples of repetitive cases of movement of a pendulum or a solar system – no concrete succession of events can be described or explained by any one law let alone any trend.

One may however argue (as Mill did) that trends can sometimes be deduced from "true natural laws" in which case we have sufficient ground for prediction. And a law may assert that under some special circumstances (initial conditions) certain trends will be found; it is also possible as Popper admits, to formulate a law corresponding to the trend. He says: 'If we succeed in determining the complete or sufficient singular conditions *c* of a singular trend *t*, then we can formulate the universal law': 'Whenever there are conditions of the kind *c* there will be a trend of the kind *t*. The idea of such a law is unobjectionable from the

---

[13] Same page (iii) and also the footnote 1 is important where Popper maintains that confirmation by many facts is easy, but corroboration is important, a theory is corroborated only if we are unable to find refuting facts. One of the examples criticized in the procedure of Toynbee's allegedly empirical investigation into the lifecycle of what he calls 'species civilization', but by the latter he means entities that conform to his belief in life cycles.

[14] Ibid. p. 115.

logical point of view; but it is very different from ... an absolute trend or a law of mathematical sequence ... Besides how could we determine that our conditions are sufficient?'[15] In the same footnote mentioned above Popper says, 'In order to test such a law we have to try hard to produce conditions under which it does not hold; to this end we must try to show that conditions of the kind *c* are insufficient, and that even in their presence a trend like *t* does not always occur'.

Popper maintains that the historicists overlook 'the dependence of trends on initial conditions' and therefore the latter believes in absolute (unconditional) trends. He seems to have "no quarrels with them" who realize that trends depend on conditions. Here we have two questions to consider. The first one is whether rational predictions can be based on trends and secondly the empirical question whether Marx (upholding the most "perfect and the most dangerous" kind of historicism in Popper's opinion) actually *did* confuse between trends and laws. If the answer to the first question is in the affirmative and the answer to the second one is in the negative then Popper's position against historicism in general and Marxism in particular do not stand.

Let us start with the first one first: it can be argued[16] 'that statisticians such as actuaries, sufficiently cautious to make large profits for insurance companies, do predict on the basis of trends'. Decision theory also is based on an assumption of rational choice and makes predictions, which may turn out to be false but nevertheless based on rational ground. Economists in similar vein make rational predictions on the basis of trends. In various scientific researches (including physics) predictions are based on trends, where no definite law is yet found. Of course these are rational because initial conditions and auxiliary hypotheses are taken into account. Even in certain cases ad hoc hypothesis also is used for a prediction. But rational prediction does not necessarily claim to be infallible or successful. Approximate predictions based on trends are frequently made in meteorology, medical sciences and also "arguably" even in "physics (increase/decrease in entropy)". Predictions on the basis of computational data are also the basis for speculation in the financial market. These are done by the stock-merchants whose sole motive is to maximize profit. So rational predictions can be based on trends for it does not presume either success or inevitability and in practice predictions are often made on the basis of mere trends.

The second question is of historical nature and therefore controversial. The question is whether Marx *actually* did confuse laws and trends. To consider this question we have to examine at least briefly his account of historical development. It is true that Marx referred to some mechanism or causal factors at

---

[15] K. Popper, *Poverty of Historicism*, Harper, 1957, p. 129.

[16] W.A. Suchting, 'Marx, Popper and 'Historicism', footnote, *Inquiry*, 15, 1971.

work in historical development. But one[17] can distinguish two distinct strands in his account of history -- one is the course of history as a whole and the other is any particular period of history. The first one is in the form of a general model of how societies are structured and how these structures change. In the Marxist account the change takes place through changes in the relative equilibrium of productive forces and productive relations at a given time. Testable consequences of such a model arise when the model is applied to particular period. Suchting gives the example of Marx's 'law of the tendency of the rate of profit to fall' as a fundamental way in which the development of the productive forces determines the specifically capitalist mode of production. Marx speaks of the bourgeois mode of production as the last antagonistic one. But it is not clear whether he thought his conception of history were applicable throughout the entire span of human history.

It is evident from the analogy with classical mechanistic model of physics and specific theories that Suchting is sympathetic with the Marxist interpretation of history (and critical of the Popperian understanding of the latter). The classical mechanistic model is one where matter as an aggregate of particles follows the Newtonian laws in its interaction. This model also does not entail any specific testable theories,[18] on the other hand is consistent with a range of different theories of specific domains. It only etches a framework for possible inquiry indicating the terms and basic sorts of items and problems to be formulated.

With these points into account it is clear that Popper's observation that no concrete sequence of events can be explained by any one law or by any one definite set of laws -- does not affect the basic Marxist position. Because what Marx claims is that there is a fundamental cause of social/historical development (namely development of productive forces) and that itself does not exclude the possibility of other causes being operative in any concrete sequence of events. Suchting thinks that 'there is no *general* objection to the idea of social development's being due to a fundamental cause and thus Popper's objection can be disposed of. In this attempt he gives the example of the case of cancer where this condition is the overriding reason for an imminent death whereas there may be other causes as well for the particular cases of death. For example a cancer patient can die in a case by a car accident or a heart attack or so on, but the fundamental cause nonetheless remains cancer in all these cases other things being equal.

---

[17] W.A. Suchting, op. cit. section VI. In this section it is argued that the mechanistic model about history as a whole is different from any particular period and the former has testable consequences *only* for a particular period.

[18] Many critics of Popper point out that the Newtonian theory of gravitation is unfalsifiable but still scientific. See Vol. I and II of *The Philosophy of Karl Popper*, ed. by Schilpp, in the Library of Living Philosopher series, particularly the articles by Lakatos and Putnam.

Can this analogical argument explain historical/social situations in the same way? Suchting's analogy can go only halfway, but it cannot explain all social situations in the same way. Popper's basic point is not that productive forces *were* not the dominant factor of social development, the significant point is that even if they *were* so long the dominant factors, we cannot predict on the basis that the trend will continue in future as well because social development does not follow any inexorable law and individuals have a lot to contribute to social development. The historical question whether productive forces in fact were the fundamental causes of social development moreover is another issue and not an uncontestable one. Therefore although one might even concede to Suchting (and Marx) that the laws of development in question are laws relating to specific historical stages, it does not follow that capitalist mode of production (the main example) is the only factor or even the main cause in bringing about social development.

About the question whether Marx thought that there could be 'absolute' 'unconditional' trends, we should distinguish between a concrete situation and how it is explained by theoretically constructing it from the elements and the assumed laws of their behaviour. He does not take what he calls laws as unconditionally operative. That means initial conditions are also necessary in his account for a prediction. The example of the law of accumulation of wealth shows that 'an exact analysis of the process' of capital accumulation 'demands that we should for a time, disregard all phenomena that hide the play of its inner mechanism'[19] The phenomenon of polarization of wealth (in fewer hands) and poverty (in more and more cases) is called by Marx as 'the absolute general law of capitalist accumulation'. Popper cites this[20] as a case of unconditional prediction ('prophecy'), but overlooks that part 'Like all other laws it is modified in it's working by many other circumstances, the analysis of which does not concern us here'.[21] But despite this remark, it is clear that Marx used 'absolute' in the context of a law to mean something like "fundamental" similar to the use of law of gravitation as absolute or fundamental in the Newtonian mechanics. The law of gravitation is "absolute" in the sense that all bodies (unconditionally) have a tendency to acceleration towards other bodies. But the tendencies can be affected by the effects of other factors. In fact many physical laws (and possibly all) are hardly ever-operative independent of countervailing factors.

Therefore one who is defensive of Marx's use of social laws as "absolute" can argue that being fundamental does not exclude the possibility of interference by countervailing factors and secondary laws even in physical sciences. In a similar vein Marx notes 'the countervailing tendencies, which have a continuous decentralizing effect alongside the centripetal one' while he speaks of the

[19] *Capital*, I, p. 565.
[20] *Open Society and its Enemies*, chap. II
[21] *Capital* I, p. 644.

process of 'centralization of existing capital in a few hands and a deprivation of many of their capital'. Suchting points out another significant and overlooked point in relation to the Marxist law of falling rate of profit which Marx considers to be a 'law of the tendency' of the rate of profit and delineates in details the "Counteracting Influences" that affect the operation of the law.[22] Although there are scattered passages that mention 'free-will' as one of the human qualities, the position remains basically of upholding social determinism. It is evident from his stand on the proletariat as well as bourgeoisie. The important question to him is not what particular proletarians at particular moments consider as their aim but "what the *proletariat is*". 'Its goal and its historical action are obviously and irrevocably prescribed for it' by its own position as well as by the particular social situation. Similarly placed are the bourgeoisie - obliged to do what they are compelled to do.

To come back to our original enquiry whether Marx did actually confuse between laws and trends, it is reasonable to argue that although there are references here and there to countervailing factors or even free will - the overwhelming reason still remains to think that he regards social trends as absolute and deterministic. Frequent use of terms like "necessity" "inevitable", "compelling" etc. indicates that social development at least so far has been explained in deterministic terms. The allegation that he also made unconditional predictions ("prophesies") about the future however remains contestable. Because one can make a distinction which Popper did not make between short-prediction and an absolute prediction that can make Marx's position not as deterministic as Popper and many others ascribe to it. In Popper's critique of historicism we have taken up Marx's case in particular because Marxism according to him is 'so far the purest, the most developed and the most dangerous form of historicism'.[23] But Popper's arguments against historicism in general also involve some internal problems as pointed out by many of his critics. As we have already seen, he assumes first that his historicist doctrine is characterized by prediction of future social events as its principal aim . The second: in making statements about the future it employs 'unconditional historical prophecies rather than scientific prediction'. Are these two assumptions valid? If they are not, then Popper's arguments against any deterministic view of social development, which he calls "historicism", collapse. But before examining the assumptions let us see whether the arguments against the doctrine (even if we grant Popper his assumptions) are strong enough to reject it.

The first general argument against the historicist view is that it is unscientific as it confuses trend with law. Apparently this argument is based on Popper's demarcation criterion of falsifiability as a mark of a scientific statement.

---

[22] *Capital* III, ch. i. xiv.

[23] *Open Society*, II, p. 81.

Popper's argument runs like this: 'the evolution of life on earth, or human society is a unique historical process'. Universal laws are about some 'unvarying order which should be tested by some new instances of the case' ("corroboration" in his methodology). His argument against historicism is that since the universe of discourse in this case contains only one element, namely the actual historical sequence, the "historicist law" must be untestable and therefore "unacceptable to science". To be fair to his critics Popper anticipates two possible objections to his argument. One: the evolutionary process is *not* unique, and the same process can be observed in many cases. Two: even if it is unique, "a trend or tendency" or direction can be discerned and that "we may formulate a hypothesis which states this trend and that 'we may formulate a hypothesis which states this trend, and test this hypothesis by future experience'.

Peter Urbach[24] argued that Popper did not succeed in replying to these two possible objections and therefore could not refute historicism as the latter claims. Urbach's presentation of Popper's view is like this: although there are parallel between certain types of historical events 'all these instances of repetition involve circumstances which are vastly dissimilar, and which may exert an important influence upon further developments'.[25] From this Popper concludes that from the apparent repetition of an historical development, we *cannot* validly expect that it will continue. Urbach finds this argument invalid because Popper's ground for rejection of historicist theories is that we cannot validly infer the historicist theory from observations of particular historical developments. He adds 'It is, however, clear that this remark does nothing to establish that historicist theories must be *untestable* and it is odd, to say the least, to find Popper objecting to a theory on the ground that *it is not proved from facts*. After all, the egregious feature of his philosophy is that the credit is given just to those theories which transcend the given facts and are as testable as possible.'[26]

Now here is a misunderstanding of Popper's ground against historicism (or determinism of any sort). His criterion of falsifiability as a demarcation criterion between scientific and unscientific demands that a statement (theory) is scientific if it is testable (falsifiable in principle) by some observation statement. He never objects to a theory as Urbach alleges if the theory 'is not proved from facts'. In fact no statements except trivial truths can be *proved* in his view. Indeed Popper speaks of corroboration as a further methodological requirement for theory acceptance, but that is the theory fulfills the requirement of testability/falsifiability. Moreover his objection against historicist theories is based on fundamental logical principles, if only *after* corroboration one accepts his description of social development as a *"unique historical process"*, "a singular historical statement". And as "unique", "singular" sequence this can be

---

[24] Ibid.

[25] K. Popper, *Poverty of Historicism,* pp. 108-109.

[26] Peter Urbach , p. 119. The second emphasis is added.

expressed *only* by existential proposition in the form 'for some x, x is a historical event'. In other words the statement asserts that 'there exists such and such a historical sequence of events'. And as it is just an existential statement, no possible state of affair can falsify it, therefore it is not testable (falsifiable) and therefore not scientific. So it is wrong to argue that Popper rejects historicist theories because they are unprovable; his ground is rather that these are not genuine laws, therefore not universal propositions. Only the latter are falsifiable, existential propositions are not falsifiable; therefore, from a singular description of events we cannot infer a future sequence of similar events.

About the other part of Popper's rebuttal however, Urbach is right to point out that Popper's crucial distinction between trend and law, although acceptable in principle, is false as a descriptive claim. I have already discussed that in connection with Suchting's account of empirical success of many predictions on the basis of trends that it is not correct to say that predictions based on laws are valid. Urbach also refers to Suchting's example (I have argued only in more details with some more examples) to argue that predictions can be validly made also on the basis of trends some of which may later turn out to be laws or based on laws. There is another argument against historicist theory besides its being "unscientific". This argument is meant to establish that the historicist laws are false. Popper alleges that no concrete sequence or succession of events except cases like the movement of a pendulum or a solar system can be described or explained by any single law or by any definite set of laws. The argument assumes that (1) historicist law claims to explain social events by the help of one single law disregarding other laws and that (2) human society is crucially different from pendulum or solar system in their repetitive mechanism. About (1) we have already seen that explanation of a social event is never done by the help of one single law. And the variety of initial conditions which is crucial for laws are not totally ignored even by Marx. It seems therefore that perhaps Popper presumes that *if a sequence of events is determined by initial conditions and a set of true laws then there can be no single law - true universal statement which can describe that sequence.*

But this seems to be not the case. An example (suggested by Urbach but I made variations in details) would suffice to show that the metabolic processes of all animals are determined by physical and chemical laws together with some relevant initial conditions like the climatic condition, kind of food available, the kind of habitat (sea-level or mountain, forest or desert etc.) and so on. But then we can have numerous true, strictly universal laws describing these situations in the animal world. 'All animals in the Himalayas have more fur than the animals in the plains' or 'the cows in eastern India are very weak' - are examples of such true, strictly universal statements. But it is clear that Popper would not regard these "strictly universal" statements as genuine laws and rather consider them as "logically insufficient and intuitively inadequate". His account of laws of nature is much stronger than merely "strict universal"; it includes an element of "necessity". Laws of nature are necessary not in the sense of logical necessity - but in a little weaker sense i.e., in the sense of physical necessity. We shall come back to this later but in connection with historicism he condemns its pretension to have found the "inexorable laws of succession". But then even physical laws are *not* necessary in the sense of being true also in other possible worlds varying in initial conditions. Therefore if we *can* find merely true universal statements about social development, they are on the same footing as physical theories which occupy a conjectural status in Popperian methodology.

About the point whether solar system and pendulum (Popper's example of systems where a single law and the same set of laws are operative) are crucially different from human society, Popper's account is not very clear except that the former is repetitive whereas human society is subject to so many influences that its development is spontaneous and not predictable.[27] But even the example of pendulum and solar system are subject to other conditions. The working of the pendulum is contingent upon mechanical and also chemical laws. Solar system again may be influenced by other galaxies - although it may take two million years. To come back to the issue of validity of Popper's two assumptions in this context. The first assumption is that historicist doctrine is characterized by prediction of future social events as its principal aim. I think Popper's emphasis here is misplaced and therefore target of many critical assaults. It can be argued instead that *explanation* of social phenomena in terms of some determinist law characterizes historicism rather than *prediction* ('prophecy' in Popper's terminology) of the future events although there are examples of social predictions also in some cases.

Indeed Popper provides us with an account of historical explanation. In his account all causal explanation of a singular event can be said to be *historical* in so far as the "cause" in those cases are described by singular initial conditions. Universal law and specific events are together necessary for any causal explanation. And in Popper's view historical events are *unique*. But historical explanation of typical events must necessarily be treated as typical "Disentanglement of causal threads" and the description of the 'accidental manner in which these threads are interwoven' -- are the two necessary tasks of history.[28] Popper thinks these two aspects supplement each other - an event may be "typical" from the standpoint of its causal explanation but "unique" at another time. He rightly points out that historical theories are not like physical theories. 'The attempt to follow causal chains in the remote past would not help in the least, for every concrete effect with which we might start has a great number of different partial causes; that is to say, initial conditions are very complex, and

[27] Urbach argues that Popper's argument to show the difference begs the question. See, op. cit., p. 123-124.

[28] *Poverty of Historicism*, p. 147.

most of them have little interest for us'.[29] He recommends a *"preconceived selective point of view"*" for any history. These "approaches" or "points of view" obviously are selective and therefore not testable.

So it is better to call them *historical interpretations* rather than historical hypotheses. The historicist view in his opinion is myopic because 'they do not see that there is necessarily a *plurality of interpretations* ...' 'Instead they present them as doctrines or theories, asserting that "all history is the history of class struggle" '.[30] We should be aware that in history one adopts a particular point of view - it is only one among many and it is possible to have other plausible interpretations. If these were the *only* criticisms of historicism, one could be in agreement with Popper. But he presented social sciences as if their principal aim is to make predictions. First of all, there is a wide range of theories of different social sciences and each science differs in their assumptions and aims. Although there is some amount of predictive element on the basis of these theories, it would be simplistic to hold either that prediction is the principal aim of social theories or that even if some predictions are made, these are done in the same way as is done in mechanistic physics or/and presumed to be inevitable.

It is obvious that social scientific theories aim principally to explain social phenomena and often resort to the method of retrospect, i.e. from effect to cause and not from cause to effect as Popper presumes. Indeed explanations in social scientific theories involve causal processes but mostly these are *not* causal explanations in the strict sense that first a cause (a prior event) happens and on that basis prediction of an effect (a later event) is done, but on the basis of existing trend it is possible to predict some future phenomenon. For example from our past experiences that political instability or a natural calamity causes uncertainty in the financial market, we can rationally predict that there will be a dip in the stock market if similar condition prevails. This prediction is based more on a rational explanation than causal explanation.

Any anti-determinist theory of social change therefore needs to provide with a non-causal account of it. Moreover if one regards (as Popper does) social laws as not "an inexorable process" and considers that individuals have an important role in bringing about change -- then a full-fledged theory of social change is required. Popper realized the need and proposed his theory of "piecemeal engineering" (as contrasted to the holistic "social engineering") which, I shall argue, suffers from many difficulties. It is clear that deterministic theories do not have the burden so much as the non-deterministic theories because the former regards change/future events as inevitable effects of some prior events. But the non-determinists require answering the questions why and how the changes occur.

---

[29] Ibid., p. 150.

[30] Both quotations are from p. 151.

At this juncture we need to differentiate between two questions: (1) Should we know social processes only for knowledge's sake? Or (2) Should we know social processes with a view to *change* it? Different social sciences assume the affirmative answer to the first question as natural sciences do but the second question is also very significant for them. Because knowledge of social processes aims often to change their course by individual interventions. In a sense knowledge of physical reality also helps to change or at least modify its course for the better. But "pure science" is at least possible in natural science, but in that sense it is not possible in any social science.

Popper's piecemeal engineering is an anti-holistic alternative to understand social processes. As contrasted to holism (a view that tries to explain social changes by social wholes, i.e. entities like states, governments, economic system etc.) it is an attempt to solve a suitably small problem by small adjustments. In other words it tries to avoid changes which are too complex and have too big a scope. The unintended consequences of such changes are then tried to be minimized. It can be done by two possible ways; either by changing as few variables as possible, or by making quantitatively small changes. Popper recognizes both but does not clearly separate them. It is also possible to make another distinction[31] between piecemeal engineering as a method of obtaining scientific knowledge and as a method of changing society. This distinction is crucial if we consider the two interpretations of piecemeal engineering, namely changing few things and making small changes. Popper's strategy for understanding social phenomena and their interrelations is to change as few factors as possible. Ideally if we change only one variable and observe the consequences, that way we will be able to know the effects of the change we have introduced. Manipulation of several factors at the same time will not help much since it will be very difficult then to know exactly the role of these various factors.

This strategy has both a cognitive and a moral ground. Cognitively it will enlighten us to know the role of a particular variable and thus the inner dynamics of social events with their complex causal processes would be to some extent clearer. And the moral aspect of such a strategy is to minimize unexpected and undesirable consequences. After all minimization of suffering is a greater aim of public policy than maximization of happiness - the latter being dependent on many private factors.[32]

---

[31] P.D. Shaw has made such a distinction. See his 'Popper, Historicism and the Remaking of Society', *Philosophy of the Social Sciences*, I, pp. 299-308.

[32] 'I proposed that, in the main, happiness should be, and that it can only be, left to private initiative, while the alleviation of avoidable suffering is a problem of public policy', K. Popper, 'The Moral Responsibility of the Scientist', *The Myth of the Framework,* ed. by M. Notturno.

This method seems to be dependable because to observe the effect of an event x on y, the experiment is designed in such a way that the effect of all other factors except x is ascertained as much as possible. The strategy is similar to the use of control variables (to minimize the effects of all outside factors), randomization etc. to isolate the separate effect of one variable to another. But the situation in social sciences is much more complex than classical experimental studies. It is not possible in many cases to manipulate one variable without disturbing others or else no manipulation is possible at all. These problems are however not crucial for Popper because the dynamics of social processes are much more complex and he regards piecemeal engineering as an interventionist method and therefore inferences based on statistical data would not either be endorsed or rejected by the method. Some critics however compared piecemeal engineering favourably with some standard statistical methods of analysis used by sociologists and economists.[33] But I do not think that this move is either helpful or necessary.

It is not helpful because statistical method even if able to establish some statistical laws (correlations) cannot work on so many variables as social events, and it is not necessary as estimating prediction is *not* the important task for piecemeal engineering as a method for social sciences. Because the future is open. It is not predetermined and thus cannot be predicted - except by accident. The possibilities that lie in the future are infinite. 'When I say, "it is our duty to remain optimists", this includes not only the openness of the future but also that which all of us contribute to it by everything we do: we are all responsible for what the future holds in store'.[34] So, Popper concludes that it is our duty 'not to prophesy evil, but rather to fight for a better world'. From the above quotation it is clear that piecemeal engineering need not have the approval and support of any statistical methods (Irzik also compares it with "path analysis") to estimate causal parameters of a postulated causal system, that is to measure the causal impact of one variable upon another in that system.

Piecemeal engineering as a method (of changing society) recommends small adjustments as opposed to undertaking a programme of drastic changes. The rationale for such move is clear -- to avoid unforeseen consequences and harms to people as much as possible in any social experimentation. This is again in tune with the indeterministic view of society. If the future is open and indeterminate, it is better to take less risk by introducing small changes (in scope and

[33] Gurol Irzik finds such association with the statistical method of "Regression" analysis which is a method for estimating the relationship between a dependent variable (say period of T.V. watching and one or more independent variables (say education and income). Its standard use is to predict one variable from a set of other variables. 'Popper's piecemeal Engineering: What is good for Science is not always Good for Society', *British. Journal of Philosophy of Science*, 36, 1985.

[34] Ibid., Introduction, xiii.

magnitude) than take a greater risk, which may involve greater harm to people. Reform therefore involves less danger than revolutionary changes.

But unfortunately this aspect of the methodology involves many problems. The internal problem is that 'little changes as opposed to big ones is not a desideratum for social scientific methodology, but *a restriction imposed from outside* on the scientist's (or the reformer's for that matter) work independently of his scientific commitments'.[35] The reason for taking this as "a restriction" upon the methodology is that the considerations taken into account here, namely "smaller changes bring smaller harm" is a sociological consideration and does not follow from the scientific commitments. Magnitude and scope of the adjustments varies from case to case. These things cannot be decided by any a priori desideratum, especially if individuals are ascribed a social reality independent of society, community etc. as Popper's methodological individualism presumes. Because how much adjustments would bring desirable change depends on particular set of individuals concerned. We can draw an example from psychotherapy methods in this context. In some cases slow and gradual treatment is more effective but in some other, sudden and drastic measure (shock therapy for instance) is more helpful in curing the patient. Similarly some social institutions for example caste-based marriage and untouchability can be changed by slow and small adjustments while witch-hunting, child-marriage or Satee (the ancient Indian custom of bride sharing the pyre of her dead husband) should be changed by ruthless and uncompromising methods.

Small changes are not suitable also where the causal association between the events is not very strong. In these cases small manipulations are not of much help because a great amount of change can only bring a noticeable change. For example the correlation between growth of industry and inflation rate is weak, slight changes in these cases cannot bring appreciable difference or/and still show many counter-examples.

Another problem inherent in the piecemeal engineering is the problem of thresholds - well known in psychology of perception where it is observed that gradual changes in the stimulus do not bring any noticeable change in the perception until a certain amount (a threshold) is attained when it brings a substantial change in the perception. In a similar way small increments in the independent variable do not produce any change up to a certain point, but a sudden change in the dependent variable occurs after a certain threshold is reached. Popper's belief that quantitatively small changes can bring only small change and thus less harm to people is also not well grounded. The assumption that the magnitude of the causal event is proportional to that of the effect is also unwarranted. The single piece of information (known only to the British intelligence) that Jinnah the Muslim leader who was the proponent of the "two

---

[35] Irzik, *op cit.*, p. 5.

nation" theory and wanted partition of India –is terminally ill could have made an enormous change in the Indian political scenario in 1947. The sudden death of Mir Mardan (of Nawab Sirajuddaullah's army) from a canon ball at the Plassey changed the whole course of the battle and history of not only Bengal but of whole India. It was vital for the defeat of the army of the Nawab and thus is instrumental in ushering the British rule first in Bengal and then in the whole of the Indian subcontinent.

Moreover the univariate (with single variable changed) model is not always the most effective. Changing few (and ideally one) variables does not always bring the desired result. Let us take the example of changing life style. If we take a single independent variable income for example, as the cause of improvement of life style (the dependent variable) and increase it to expect the result, more often than not it does not happen. A daily labourer may indulge more in drinking and other habits if the wage is increased and not necessarily have a better life. The utility value of anything depends also on the capacity of the individual concerned. Unless other independent variables such as education, information opportunities etc. are changed, the desired effect (an improvement in the life style) cannot occur. Therefore piecemeal reforming by small adjustments cannot bring any appreciable change in most cases unless aided by changing other independent variables as well.

It is evident from the above discussion that there is a basic difference between the method of gaining scientific knowledge and that of changing social institutions. We have seen that changing *one* independent variable at a time helps gaining knowledge about co-relations among events whereas that method is not effective in bringing any appreciable change in social institutions. In the latter case the situation is much more complex because of participation of individuals and their free interrelations with each other and with the institutions. Piecemeal engineering a method of social change as envisaged by Popper fails to encompass all kinds of situation. It assumes or at least in principle requires that the society is more or less free. But suppose a society is *not free*, either in the sense of having a totalitarian political regime or in the sense of being dominated by fundamentalist religious groups and opposed to humanistic norms; in that kind of social situation making minor reforms will either be not allowed or would be allowed *only* if these are in favour of perpetuating the condition of unfreedom.

Neither of these alternatives are consistent with Popper's original reason for favouring gradual and small adjustments, namely to protect people from harm and undesired consequences. Therefore in such situations drastic and large scale change is the only possible move to improve the social situation and protect peoples rights. Here I assume that changing a social situation implies changing it *only* for the better. Similar baffling situation is faced by Gandhian non-violent method when hypothetically it would face a totalitarian power like the Nazis or the Fascists. If the movement is to change a social situation, it should be effective, and non-violent method cannot succeed if there is no basic freedom in that society. Moreover this method is ironically also applicable to a totalitarian society to maintain their status quo by not encouraging large-scale change. This would be self-defeating for Popper's "open-society".

To conclude we can say that social cause and effect are not always proportional so that small changes bring smaller harm and large-scale changes bring undesired consequences. Secondly even if we accept that the future is open, surely some sort of lessons we learn from history. If trial and error is a process through which we *know* the world then that method is also applicable to our knowledge of social phenomena. Indeed we make mistakes and often 'do not learn from history'. (The standard example is of Hitler's army making a similar mistake as Napoleon's army in invading Russia) – but more often than not we learn from past experiences *not* to make too many mistakes and take a cautious stand to avoid similar harms that hurt us in the past. This kind of precautionary move does not presume social determinism nor deny that the future of social phenomena is indeterminate.

## *The Open Universe and Propensity*

The criticism of the thesis of determinism is crucial for reality of human freedom, creativity, and moral responsibility as well as for the question of human rationality. Popper is one of the main critics of physical determinism the truth of which in his view renders our notion of ourselves as rational agent completely meaningless. In his critique of the nightmare of the physical determinist' two concepts are very crucial, (i) the concept of a physically closed system and (ii) an open system. A deterministic physical clockwork mechanism (such as Newtonian 'giant clock' world) is completely self-contained, in the sense that it is physically pre-determined leaving no room for outside intervention. He says 'By a physically closed system I mean a set or system of physical entities such as atoms or elementary particles or physical forces or fields of forces, which interact with each other - and only with each other - in accordance with definite laws of interaction that do not leave any room for interaction with or interference by anything outside that closed set of physical entities. It is this closure of the system that creates the deterministic nightmare'.[36] The reason for calling determinism a nightmare is that it regards the whole world as a huge automation reducing us to the status of 'little cogwheels or at best sub-automata'.

As open system by contrast is one that is at least partially open to outside influence and *not* a predetermined complete system which contains *all* the factors for the system to function. Thus creativity, free and rational action requires at least an open and indeterministic universe. However it is evident that

[36] K. Popper, *Objective Knowledge*, p. 219.

mere indeterminism is not enough for providing an adequate account of human freedom and rationality. Popper realized that indeterminism can only be a necessary condition for human freedom and tried to provide with some extra conditions for an adequate account of freedom. I shall try below to examine the arguments against determinism, for an indeterministic open universe and also the ground why indeterminism is not enough.

## *Against Determinism*

As noted earlier the general doctrine of determinism can be understood in two ways, (1) The thesis that in principle all events are predictable which is named by Popper as 'scientific determinism' (2) This distinction as emphasized by Popper is subject to lot of criticisms.[37] We shall therefore examine the arguments that are forwarded against determinism in general and also the position (usually known as compatibilism) those claims to reconcile determinism with individual freedom which Popper rejected.

The classical example of a deterministic position is one that is presented by Laplace and his demon. According to this view the world is made of particles acting upon one another according to Newtonian dynamics, and that a complete and precise knowledge of the initial state of the world system at any given point of time would be sufficient for the deduction of its state at any other instant. We know that if the complete initial conditions, i.e., the positions, masses, velocities and directions of the movement of all its particles are given -- then the "state" of a Newtonian system is given. It is clear that only a superhuman intelligence is only capable of ascertaining the *complete set* of initial conditions of the world system at any given point of time. Hence the concept of the demon as Laplace has it, with the laws of nature (equations of mechanics) aided by the initial conditions *it is possible to deduce all future states of the world system*. Laplace's idea is to show that given the laws of nature as known, the future of the world is predetermined in any instant of its past. The significance of this view is that the demon is not God, only a super-scientist – that is an all – perfect image of a human scientist. The only difference being his knowledge of all the initial conditions which the human scientist cannot achieve. The idealized version of a scientist thus (as envisaged by Laplace) can solve the great problem of the stability of our solar system if the number of planets is small. Indeed the human scientist cannot obtain absolutely exact initial conditions, but the degree of precision can be improved in the way of approximation.

---

[37] Peter Clark, 'Popper on Determinism', *Karl Popper: Philosophy and Problems*, ed., Anthony O'Hear. Clark argues that Popper's account does not distinguish clearly between the notion of physical systems and that of 'possibility' which is a logical notion.

Popper calls this kind of determinism as "prima facie determinism" and not only Newton's theory exemplifies this kind of determinism but certain features of Maxwell's or Einstein's theories also fulfill similar features) 'in contradistinction to other theories such as thermo-dynamics or statistical mechanics, or quantum theory and perhaps also the theory of genes'.[38] A physical theory is *prima facie* deterministic 'if and only if it allows us to deduce from a *mathematically exact* description of the initial state of a closed physical system which is described in terms of the theory, the description, *with any stipulated finite degree of precision*, of the state of the system at any given future instant for time'.[39] Even though the initial conditions are mathematically exact, this definition does not claim mathematically exact predictions since only approximate methods are known to solve problems even of Newton's mechanics (if the problem concerns gravitational interaction of more than two bodies).

Another argument against determinism is given by Heisenberg on a different level. It exploits the notion of indeterminacy of the standard of measurements. Determinism is not acceptable in view of the fact that due to the interference of the process of measuring with the state of the measured system, there are definite limits to the possible precision of our knowledge of the initial conditions and therefore to the predictions which can be calculated from them. This argument rejects the unlimited power vested on the demon by Laplace. It is based upon the implicit assumption that determinism entails *predictability from within* with any degree of precision.

Popper however does not accept the idea that quantum indeterminism is due to the interference of the measuring process with the object measured. In fact he tried to show that certain features of quantum physics are also found in classical physics.[40] Two principal arguments against any version of scientific determinism are based respectively on the approximate nature of all scientific knowledge and on the asymmetry between the past and the future also known as the irreversibility or arrow of time.

## *(a) Approximate nature of scientific knowledge*

The world or nature is too complex. As is well known, to Popper theories are rational attempts ('nets') of our own making by which we can try to capture the real world. Even if these are approximations to reality - theories (conjectures) can never assert that the world is determined i.e., known fully with its causal processes. From the constant changes and refutations of theories, the

[38] K. Popper, *The Open Universe*, pp. 30-31.
[39] Ibid., p. 31.
[40] K. Popper, 'Introduction', *Quantum Theory and the Schism in Physics* and also *Open Universe*, p. 35.

hypothetical nature of our cognitive system is clear. Creative work, intellectual achievements can never be fully accounted for by any scientific theory.

Moreover the uniqueness of the world at any point cannot be fully explained in terms of universal theories. However perfect the theories are, they are approximations, so attempts to describe nature in terms of theory will lead to an infinite sequence of approximation, because our theories are attempts to rationalize nature. As knowledge is conjectural, we can at most claim that the past and the present state of knowledge can be to some extent ascertained. But in no circumstances we can claim knowledge of any future state of events. Therefore however well we know the present state of affairs, we can never predict on that basis which direction our growth of knowledge will go in future.

### *(b) Asymmetry between the past and the future*

There are good scientific reasons, stemming from a *prima facie* deterministic theory namely the special theory of relativity, 'which support the commonsense view of the openness of the future'.[41] In physical terms, this asymmetry is established by the fact that from any place in the past, a physical causal chain (for example a light or a sound signal) can reach any place in the "future" but from no place in the future can such an effect be exerted on any place in the past. As a consequence, the future becomes "open" to us in the sense that it cannot be fully predicted by us but the past is closed, nothing else can influence it to change. This is the asymmetry between the past and the future. The following figure (slightly different from Minkowski's and Popper's versions) can help us to understand the asymmetry and the difficulty of predicting future phenomenon from the past.

[41] For the proof, see, *Open Universe,* pp. 58-61.

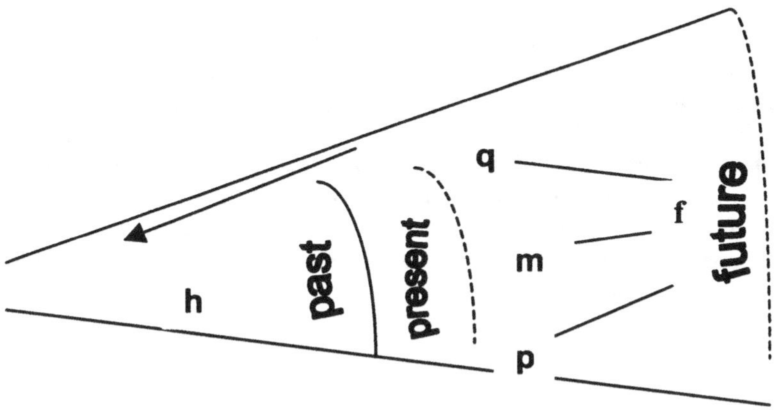


From h if we have to predict a future phenomenon f we see that only the factors within the past cone h are (only ideally) known to us. But the past cone of f includes many factors (say p, m, q) that are outside the past cone of h but can influence the occurrence of f. From h however we cannot know any thing like p, q, and m. Therefore from the present point h we cannot predict any future event that is anything that falls beyond h. Even the Laplacian demon can predict only if the number of initial conditions is limited. As Hadamare has shown, the problem of the stability of the solar system cannot be fully calculated on the basis of any universal theory (e.g. gravitation) and measurement of the initial conditions however precise they may be. So the question of deciding about some specified state of a many body system whose initial conditions are given with any finite degree of precision remains a physically undecidable question.

But the question still remains what is meant by the openness of the future? Does it mean uniqueness, newness or failure of past events to account for future state? Inability to predict does not entail lack of causal connection between the past and the future? In Popper's own account theories are falsifiable and only attempts to rationalize - then it may very well be the case that a theory overlooks a certain possibility of a causal chain. A possibility cannot be new only if it is not known to us. Indeed we *may not* know many possibilities within the past cone. Again it is not clear why even if the future is "contained" in the past it has to be "redundant" and "superfluous".[42] For example a computer's functioning depends on the hardware, software, programming and also the operator but that does not

[42] K. Popper, *Open Universe*, p. 91.

mean that the computer responses are redundant and superfluous as they are "contained" in a sense in these factors belonging to the past. A further difficulty appears in Popper's criticism of Einstein's interpretation of Minkowski space-time of special relativity as leading to a four-dimensional block universe.[43] It can be argued that a four dimensional block universe is not committed necessarily to determinism, one could have such a model of an indeterministic universe ... 'all that would require would be that the events comprising a suitably chosen time-slice of that Universe did not fix together with the appropriate laws of nature for that universe, the events associated with later times'.[44] That means the block-universe models 'may have implications for the reality of time or change', or to be precise, for the reality of tense, but that does not have anything to do with the truth or validity of determinism.

## *Indeterminism and Freedom*

One of the main arguments used against metaphysical determinism is known as Lande's blade. His argument aims to show that we must accept probabilities of single events as fundamental and as irreplaceable by any statement except by other probability statements. He shows that even if we combine a *prima facie* deterministic theory with statistical assumptions concerning initial conditions, we only get an infinite regress, and an interpretation which sticks to this assumption is surely to become untestable and hence metaphysical. The importance of Lande's Blade[45] is that it shows the emptiness of the old deterministic "explanation" according to which many small causes or "errors" will produce the random results. It suggests that any deterministic explanation of statistical behaviour allow 'considerations of probability and also propensity to slip, unnoticed, into their assumption'. They assume that *uncontrolled initial conditions* are always random (the chaos theory was unknown at that time). Popper argues that this assumption may again be interpreted in a propensity sense. In this sense 'the hypothesis means that (i) the controlled experimental conditions do not absolutely fix the initial conditions but leave them a certain amount of free play and that (ii) each of the possibilities thus left open to the initial conditions will be realized with a certain propensity or probability'.[46] In brief Lande's example and argument has shown that *even in deterministic*

[43] This argument is due to Peter Clarke, 'Popper on Determinism', in *Karl Popper: Philosophy and Problems*, ed., A. O'Hear.

[44] Ibid., p. 158.

[45] Alfred Lande, 'Probability in Classical and Quantum theory', *Scientific papers presented to Max Born, 1953* and *Foundations of Quantum Theory*, pp. 3ff.

[46] K. Popper, *Open Universe*, p. 102.

*framework some element of randomness creeps in for any explanation of statistical behavior.*

The significance of the propensity interpretation is that an objectivist interpretation of probability deeper than a frequency interpretation is needed to show an intimate relation between propensity and indeterminacy. This interpretation also helps us to understand why indeterminism is not enough for human freedom. Although indeterminism is necessary it is not sufficient for creativity and freedom to flourish which need some more conditions. The physical universe must be causally open to influences from outside, not only mental activity but also abstract influences such as arguments and critical discussions. Later Popper[47] emphasizes that just as the propensity interpretation is not of much value without indeterminism so is indeterminism without full-fledged propensities. This will be clear if we try to analyze the doctrines of determinism and indeterminism.

Determinism is the doctrine that generalizes the fact that many aspects of our life show repetitions and strong regularities, regularities, which are beyond our power to change. The problem however remains how to explain the streaks of irregularities that permeate the regularity, for the unending novelty whose irregular (random) but not quite frequent appearance is almost taken for granted and for those regularities that we seem to alter. The doctrine of indeterminism on the other hand can make the "clouds" comprehensible but not the "clocks". Popper finds an asymmetry here. The significant theme of his account is that indeterminism does not suggest that no event is determined in advance, only that there are some that is not.[48] Hume's celebrated use of "law" and the compatibility thesis asserts that there may be chance-events at the micro-level as well as causal determinacy at the micro-level. It denies however any middle possibilities. In endorsing Popper's view of indeterminism as being the doctrine that at least *some* events are indeterminate, John Watkins comments that Hume's mistake lies in not noticing 'that a full-fledged doctrine of determinism involves *two* universal quantifiers. It says, not just that all events are determined, but that *all* events are determined in every detail (or are completely determined). Thus he took its contradictory to be, 'some events are [completely] undetermined' when he should have taken it to be, 'some events are not completely determined'. Our imagined set-up is a straightforward example of a physical situation that is only partially determined'.[49]

The principle of causal openness is required both for defeating determinism as well as for accounting for human freedom and creativity. Cartesian dualism and any other version of it were subject to heavy onslaught due to the

---

[47] K. Popper, *The World of Propensity.*

[48] K. Popper, *Open Universe*, p. 28.

[49] J.W.N. Watkins, 'Three Views Concerning Human Freedom', *Nature and Conduct*, ed. by R. S. Peters.

presumption that it is impossible for a man's mind to exercise a degree of control over his body (which Popper calls "plastic control"). The objection implies that: (a) A man's body is part of the physical world; if it were open to quasi-causal interference of an extra-physical kind, then at least that bit of the physical world would remain physicalistically indeterminate. (b) Something immaterial (human mind) cannot act on something material. With the break down of physical determinism the first objection dissolves and both the classical and modern science have principles that endorses the magnetic or animating effect of immaterial phenomena on material things. The idea of field of force acting on things does not even need any material medium (with the abandonment of the ether) for propagation of energy. Thus the concept of immaterial substance acting on material things has been possible by scientific developments. It can also show that the philosophical requirement of causal openness for human freedom (in addition to indeterminism) is consistent with existing scientific theories.

Some critics however point out that the doctrine of indeterminism may not be as innocuous as it seems. According to John Earman[50], it does not imply that indeterminists are not free to espouse some kind of partial determinism and to explain clocks as more or less deterministic systems more or less immune from external influence. The second argument in favour of indeterminism along with propensity interpretation is based on some ability to explain statistical regularity[51] whereas determinism cannot explain irregularity except that these are unrelated effects. This claim if valid[52] would give the doctrine of indeterminism a clear advantage for its problem-solving capacity over determinism.

### *Why Indeterminism is not enough for Freedom*

The question still remains to be seen why even indeterminism (if determinism is regarded as unconvincing) is not enough for human freedom and creativity? It is clear now that from the Laplacian idea of a causally closed and deterministic world - we can never argue for genuine human freedom and creativity. But from Maxwell's failure to reduce electricity and magnetism to Newtonian mechanics (by the mechanical concept of ether) -- there is steadily a new kind of causal openness gaining ground in physics. The still dominant trend remains nevertheless deterministic exemplified by Einstein's faith in an unified deterministic theory comprising mechanics, gravitation and electricity.

---

[50] John Earman, *A Primer on Determinism,* D. Reidel, pp. 13f.

[51] K. Popper, *The Open Universe*, section 28.

[52] D. Miller, 'Propensities and Indeterminism', *K. Popper: Philosophy and Problems*, ed. A. O'Hear. Miller is sceptical about Popper's claim in this connection.

The new kind of indeterminism introduced by quantum mechanics is different from the kind of belief in chance events, which are only due to the incompleteness of our knowledge. For example if a person is hit by a running car, it is regarded as an 'accident' or chance event, but the two causal chain processes can be reconstructed and are compatible with classical determinism. Quantum mechanics however assumes the possibility of elementary chance events which are *absolute*, i.e., not causally reducible. According to this view there are elementary physical processes which are not further reducible in terms of causal chains, but which consists of 'quantum jumps' that are absolutely unpredictable events which are not controlled or influenced by causal law, but only by probabilistic laws. Quantum mechanics regards these absolute chance events as the basic events of the physical world (Popper's World I), thus various physical processes like disintegration of an atom with consequent radioactive emission etc. are not predetermined or predictable (however great our knowledge of the initial conditions may be). All we can do about these events is to make some testable statistical predictions.

Without further details about this kind of indeterminism, the question still remains, if the physical universe is indeterministic as the Quantum physicists claim (and Popper also '... believe that even classical Newtonian mechanics is in principle indeterministic'.) - Can probabilistic laws account for human freedom? As Popper puts it very succinctly, 'what we want to understand is not only how we may act *unpredictably and in a chance like fashion*, but how we can *act deliberately and rationally'*.[53] Our deliberate and rational action needs a rational explanation - not just a causal explanation. The latter is required only when rational explanation *cannot be* given for example if somebody behaves normally (responding when asked something or so) we do not ask for a causal explanation, but if that person starts dancing in the middle of a discussion, it calls for a causal explanation like inebriated state or extreme grief or some other physiological/mental condition. In that sense the latter action is *not* free because it is still under some physical/chemical laws. But free and rational action cannot be adequately explained by referring to any deterministic law, although these are part of the physical world (speech or writing and any other action for example) - only physical laws are not enough. We need some other kind of influence from what Popper calls World 2 i.e., the world of human mind and will to bring forth free and rational action such as composition of a poem or a symphony, construction of logical proofs, a sculpture or a theory about the universe.

Indeed as part of the physical world human beings are subject to physical laws, in that sense the world is partially determinate or causally close. For example if one is deaf, it is not possible for her to compose or appreciate a symphony, but we know that there are glaring exceptions that defied natural laws to a great extent. If there is a question of influence in cases of creative activity -

[53] K. Popper, *Open Universe*, p. 126.

these influences are fundamentally of creative type. What Popper calls "causal openness" of World 1 - the physical world 2 - the world of human mind, and World 3 the world of Objective knowledge (the theorems, symphonies, computers etc. of human creation) can show adequately how influences and interactions among all these three realms can account for human freedom and creativity. No fully deterministic theories like classical physics or fully indeterministic or probabilistic view like quantum theory can explain how human beings can act freely although they are partially determined by physical laws. This is not to deny that man is part of nature but to assert that by rational and deliberate action man can transcend natural laws to some extent and on the contrary partially influence some parts of the physical world through their contribution (intellectual and creative) to the world of objective knowledge (Popper's World 3) which again can influence both the physical world and the world of human mind.

Yes, as a part of nature human being is subject to many physical laws and social traditions, but at the same time as a rational agent man can transcend many of the barriers he is born within and act to change the future. No determinist theory can satisfactorily explain free human activity.

## *Concluding Remarks*

One of the main arguments against determinism is that the future depends on the growth of theoretical knowledge and it is not possible to predict the state of theoretical knowledge from within. The argument assumes that no predictor (say a computer) can predict its own future. For if we know what we will know in a future date then the content of that knowledge (a theory) is in some sense occur to us today and not in future. Therefore, the argument goes on, we do not foresee anything that will happen tomorrow. Future growth of knowledge thus is not predictable.

It can be argued against this view that we can predict reasonably successfully that an idea or a theory will occur to someone in near future. But then it is not predicting from *within* the system, it is from without. Popper's ground against the predictive ability of future growth of theoretical knowledge is based on the example of a predicting machine. The predicting machine *cannot* predict its own future.[54] His argument uses 'the fact that every prediction of an otherwise new idea changes its character of newness. I have used the term "Oedipus-effect" as a name for the influence, upon the predicted event, of a theory, or an expectation, or a prediction, or even information used to obtain a prediction...'[55] But this

---

[54] K. Popper, *Open Universe,* pp. 64-77. He acknowledges that he owes the idea of a computing machine to A. Turing.

[55] Ibid. p. 66, fn. 1.

argument still can be countermanded by considerations such as the present state of theoretical knowledge in physics. It is a well-founded belief that in certain areas most discoveries have been already done. Then in principle at least it is possible to predict that 'there is not any fundamental discovery to be done in such and such area except some minor changes of parameters'. This kind of prediction is possible only from *within* the system (Contra Popper) and *not* from without. Moreover advance of theoretical science (at present) can reasonably bring predictable advance of technology in future.

Propensity interpretation of probability is the main pillar of Popper's argument for indeterminism. It is different from the frequency interpretation mainly in two ways: (1) it ascribes probabilities to *single events*, rather than to types of events; (2) and it relativizes probability ascriptions to the whole current *state of_the universe,* rather than to the immediate locality. For example the probability of the occurrence of 6-up with *this* throw is fixed in general not only by the physical features of the die, and by the features of the apparatus being used to throw it, and its immediate surroundings but also by the whole state of the world. The propensity (probability) is not a propensity of the die or of the apparatus but the world; it is not relative to a reference class, as the frequency theory requires. Thus the probability of a single event (say of 6-up with this throw) now lies between 0 and 1 only if the outcome of the throw of the die is not fixed in advance. The world has a propensity to develop in the way described. This position is neither rigid determinism nor absolute impossibility. Indeed the propensities in such activities as games of chance are not fundamentally propensities to yield frequencies but propensities to produce single events. There are propensities to produce frequencies but these are explained in terms of more fundamental propensities.

It is apparent that the propensity interpretation of probability is metaphysical not only because many propensities are postulated that are not empirically testable but also because the idea of necessity or of law is built into the theory.[56] But at the same time the modal element in the propensity interpretation is quite significant for without it there is no objective interpretation of single-case probability at all. But whether Popper's interpretation is objective however is a contestable issue. He introduces the concept of propensity (of an event or object) to realize alternative behaviour or states. Similarly to the case of chance, it is not possible to relate the propensity exclusively with a bare object or event without referring to the "physical situation".

---

[56] J.W.N. Watkins, 'The unity of Popper's thought' in P.A. Schilpp (ed.), sections, 2, 4. *The Philosophy of Karl Popper*, La Salle, 1974 and also 'Second Thoughts on Lande's Blade', *JICPR*,. II, 1985. D. Miller, 'Critical Rationalism and Propensities and Indeterminism' in K. Popper: *Philosophy and Problems*, ed., A. O'Hear. Both defend Popper's propensity interpretation.

Popper ultimately claims that propensity is not attached to a particle itself but to an objective physical situation in which the particle participates. The act of measurement can arrange such a situation. It implies the connection of the statistical interpretation with the result of the measurement process. According to him a personal observer's participation in measurement is a source of the subjective influence upon the result of measurement although he claims the setting up of an experiment is not related to any observer's decision but to the objective theory which underlies every scientific experiment. It seems that Popper cannot totally dismiss the participation of an observer in the measurement process, but here the participation of the subject is a result of the necessity to actualize the virtual experiment, to create the indispensable physical situation.

The very concept of propensity however is anthropomorphic, it is associated with the physical object as it not only exists but also is becoming. It resembles rather a biological or self-organizing structure than a lifeless bare object of classical physics. The propensity is associated with the process of being realized with respective frequencies. The point is that every process, which is irreversible in time, has its own individual history and future, and it needs to have a kind of self-knowledge to be able to develop in time and to self-organize. But that does not necessarily make the process subjective. It is possible in many cases to objectively record and control the process of self-knowledge.[57] The interest in human freedom and creativity motivates one to take the threat of determinism seriously. Popper's insight lies in his realization that there would be no authentic freedom in a fully determined physical world. Reconciliation of a deterministic world with human freedom as Hume proposes is outright rejected by him. Popper deploys the propensity theory to resolve a conceptual problem, namely, how to understand physical probability as it occurs in statistical mechanics. He says 'propensities can be accepted as physical realities (analogous to forces) only when determinism has been given up'.[58]

---

[57] In the light of contemporary natural science it is no longer possible to keep a sharp contrast between subjectivism and objectivism with respect to cognition of physical processes. The propensity interpretation is neither objectivist nor subjectivist in the traditional sense. M. Zabierowski in 'On the objectivity of the Popperian Interpretation of Quantum Mechanics', (*The Problem of Rationality in Science and Philosophy,* J. Minek, ed., *BSPS* 160, 1995), argues that Popper's interpretation is neither - it is rather "anthropic" as Weizaecker and Everetts quantum interpretations are.

[58] K. Popper, *Quantum Theory and Schism in Physics,* 1982, p. 105. A detailed account of the propensity theory can be found in Popper's *Realism and the Aim of Science,* 1982, Part II. Propensities are introduced to explain why in nature we almost never observe systems which have as their initial condition, states which evolve away from equilibrium. It is felt that neither in statistical mechanics nor in quantum theory does the propensity theory of probability resolve the outstanding conceptual problems of those theories. In 'A Note on Popper, Propensities and the Two Slit Experiment'. *British Journal for the Philosophy of*

Besides the conceptual problems mentioned by many critics, propensity/tendency of physical states also show both a somewhat deterministic and a becoming/changing aspect. If not only single states, but the whole physical set up is to be considered, the determinist might say that it is not far from some form of determinism explaining the occurrence of one event by referring it to not one *single* cause but the whole physical arrangement for the occurrence of it. On the other hand the concept of propensity introduces an element of change/development into the lifeless object of classical physics which can be likened to process philosophy of becoming.[59]

But even if Popper's propensity interpretation is *not* objectivist as some critics allege, does it affect his arguments for freedom? I think that it does not, for there are arguments, which he offered for human freedom and creativity. The most important of all is the argument from causal openness and the introduction of the three worlds of physical, mental and intellectual reality, which interact with each other. The notion of 'plastic control' (although not fully developed) in conjunction with propensity helps to understand the notion of freedom or open possibilities. Both in 'Of Clouds and Clocks'[60] and 'Indeterminism is not Enough'.[61] Popper elaborates how theories created by us and encoded by World I (Physical world) are sustained there as part of objective knowledge without direct intervention from us but can bring about changes in World I and by that change can influence our mind, i. e. World 2. This reciprocal influence however is not causal, human cultural attainments transcend causal explanations. We are not machines but problem-solvers responding to situations in innumerable possible ways. But mere abstract possibilities are not enough, so for actual realization he postulates active propensities, which may be utilized by us (the free agents) to bring us forward to the open future. Here we find a way out to transcend Hume's fork, namely that there is no middle way between chance and absolute necessity. It is evident now that chance is not enough for freedom and also that the latter cannot follow from absolute necessity. The third possibility is that of a world, which is governed not by necessity, but by propensity – a process of open possibilities.

---

*Science,* 36 (1987). Milne has shown that if propensity is to be seen as a property of whole experimental arrangements or set ups then since each experiment exhibits one and only one of the two aspects of quantum wave particle duality the theory cannot be used to explain the results of the two-slit experiment (as inference pattern), as an interference phenomenon of the two distinct types of arrangement.

[59] "This view of propensities allows us to see in a new light the processes that constitute our world - the world process. The world is no longer a causal machine - it can now be seen as a world of propensities, as an unfolding process of realizing possibilities and of unfolding new possibilities ... new possibilities are created possibilities that previously simply did not exit". K. Popper, *World of Propensity*, p. 18.

[60] K. Popper, *Objective Knowledge.*

[61] K. Popper, *Open Universe.*

The great contribution of quantum mechanics is that 'it teaches us that there is chance'.[62] Although it involves 'disturbing technological implications', it helps nevertheless to establish that there is a middle possibility between chance and absolute causal necessity. Watkins is right that it exemplifies Popper's conception of scientific growth. But surprisingly the latter never used the conjectural nature of philosophy of knowledge as a strong argument for indeterminism. Instead of making the ontological claim that the growth of human knowledge is intrinsically indeterministic, Popper merely made an epistemological statement that 'we cannot *predict* the future growth of our scientific knowledge'. Watkins made a much stronger claim than Popper's merely indeterministic future of scientific knowledge as an epistemological stance. 'But irrespective of how science advances it is trivially true that we *cannot* scientifically predict at a time t0 a scientific discovery that is going to occur at time t1; if we got it right the discovery would have been made at t0….'[63]. He claims [64] 'not that electrification or atomic explosions were in principle unpredictable back in 1066,say, but that they were not then causally determined.'

Indeed the indeterminism of critical rationalism is not based on the ontological chance factors introduced by the modern microphysics. Therefore it does not cognitively rely on it. Therefore it will survive even if modern microphysical theories do not. Watkins' third possibility between chance and necessity thus paves a middle path between the two where an inventive mind as well as technological innovations is both operative. It may lead us into a hitherto unknown territory of knowledge and action by free human enterprise aided by existing state of cognitive apparatus.

---

[62] J.W.N. Watkins, *Freedom After Darwin*, p. 56.
[63] Ibid., p. 57.
[64] Ibid., p. 57.

## *CHAPTER V*

# REASON, TRADITION AND FREEDOM: SOME PERSPECTIVES ON ECOLOGY

'All things living are in search of a better world'
Karl Popper, *In Search of A Better World*, p. vii

A distinction can be made between traditional knowledge-system and what is known as modern western scientific tradition. Indeed there are differences between modern science and/or non-western knowledge traditions. Although science is socially determined, that is only a necessary condition. It is true that modern science has become a source of many frustrations we can nevertheless make a distinction between science and other traditions by norms which are not very far from the rationality model yet capture the spirit of scientific enterprise in an acceptable form. Most ecological theories have virtually become accounts of battle between science and technological development on the one hand and conservation of nature and tradition on the other.

I shall try here first to analyse the issue of tradition and science as two different epistemological systems and then examine whether ecological issues are really questions of illegitimate intrusion of science into the realm of tradition as alleged by most eco-philosophers. Although Popper encourages criticism in general as a sign of a free enterprise he thinks 'all life is problem solving', as bees solve their problems human technology solves many human problems and enhance many ways of good life. 'Hostility to technology, as we find among the Greens is, therefore a foolish kind of hostility to life itself - which the Greens have unfortunately not realized'.[1] My aim is to establish that a rational theory of ecology should rather be based on an extended notion of freedom rather than vilifying scientific knowledge. The critical rationalists general position also can be endorsed for they would definitely welcome all criticisms including a critique of technology. But that does not mean all traditions are at par nor that all ecological problems are due to technological innovations and therefore the latter should be given up in favour of traditional methods. On the contrary a perfect understanding of the problems need scientific knowledge rather than eulogising *all* traditions for it is perfectly clear that all traditions are not on equal footing and often based on irrational ideology as we will see later. First we shall examine the distinction

---

[1] Karl Popper, 'Thoughts on History and Politics', in *All life is Problem Solving* , Routledge, 1999.

between tradition and science as two epistemic systems and then with the examples of ecological problems try to find out through the help of Indian cases, the issue of tradition vs. development as presented in various ecological theories. The moot point is to establish whether there is no fundamental difference between the two kinds of knowledge as most of these theories claim. As mentioned earlier most cases blame modern science for natural devastation, rather than a wrong perception of the problems.

Science seems to be the typical rational model of knowledge and universal truth. Indeed there are differences between modern science and/or non-western knowledge and traditions, but that difference is historical and contingent. In all societies and all ages, science being a culture notwithstanding, a difference can be made between degrees of validity of beliefs /knowledge. To call non-modern or/and non-western traditions nothing but culture is to deny history. Again although science is socially determined by various contingent social and cultural conditions for its actual course of research, these are nothing but necessary conditions. The question of validity of belief and knowledge is independent of its history of genesis. It is also true that the focus of scientific research often ignores human cause and prioritise prestigious areas such as space research, nuclear science or quantum mechanics more than alleviation of human suffering. Focus of research is very important because it concerns lot of funding, expertise and time. For example a common disease like malaria kills millions of people every year only in sub-Saharan Africa but much importance is *not* given to its research for an effective prophylactic which is affordable to most of the victims who are incidentally very poor.

On the other hand space research, super conductor or sophisticated military equipment get much more attention (and therefore much more funding). Moreover due to lack of transparency the actual scientific practice is often tyrannical as we examined the issue earlier. Indeed modern science has belied many of our hopes and aspirations with imbalance in its focus and priorities but that is due to science education and science policy. As a discipline science has nothing to do with its actual nature and political application. Notwithstanding all its wrong perception and values, scientific knowledge is still the only culture which expanded our horizon of knowledge of the world to an unprecedented extent. It also contains within its own tradition a self-correcting method of testability. Because it contains the tests to tell us when a set of ideas is wrong or a theory is to be rejected. So even if scientific knowledge-system can be perceived as a kind of cultural tradition, we can still make a distinction between science and other cognitive systems. The norm by which that is done is not very far from the rationality model. Yet they capture the spirit of scientific enterprise in an acceptable form by common consensus.

Tradition and reason are supposed to be two opposite notions. In this connection it is important to distinguish between tradition and orthodoxy[2]. J.N Mohanty[3] has made an excellent distinction. "Orthodoxy consists in fossilizing tradition into a lifeless, unchanging structure. Tradition, as distinguished from orthodoxy, is a living process of creation and preservation of significations". Indeed when a tradition is alive and meaningful it does not need any coercive social process for its perpetuation; adherence to a living tradition is smooth and spontaneous. Only when it loses relevance and ready response from its followers a culture becomes decadent and soon loses its natural appeal and needs coercive measures for its preservation, as found in many stages of civilizations of different societies. For example, in the Middle Ages Europe witnessed decadence of social and religious norms and thus resort to coercion and different types of persecution was rampant.

In the East, especially in India, there were periods of decadence when some tradition lost its creative growth and required orthodoxy to prove the purity of. Being more ancient than its occidental counterpart, eastern traditions were found to be more orthodox in comparison. For its long history seems to fossilize the traditions more rigidly than the comparatively less ancient western civilization. This idea has been put very succinctly by Hajime Nakamura.[4] The moot point however is to define tradition and modernity in terms of rational criticism. Moreover all traditions are not necessarily anti-rational. So we should remember that the war should not be against *all* traditions, but only against orthodoxy which defies rationality and do not respond to new situations and challenges that are not faced by earlier traditions.

The dichotomy, tradition-modernity, should be understood in its true perspective; if tradition means only obsolete irrelevant and empty adherence to customary ideas and mores, then of course it is anti-modern. On the other hand, modernity is *not* accepting everything only because of novelty and rejecting everything old and traditional. This kind of perception will help us in understanding how some kind of dialogue is possible between the modern and the traditional. Especially in connection with ancient culture and civilization like ours that has much to contribute to the general body of universal knowledge, including scientific thought. Of course it needs lot of steady and patient deliberations with an open mind and rational judgement. In this issue an interesting analysis has been done by Mohanty. In his view social and public philosophers like Gandhi, Aurobindo and Tagore have done the synthesis of these two distinct elements more fruitfully than academic philoso-

---

[2] Karl Popper has made a distinction between tradition and orthodoxy and maintains that orthodoxy 'is the death of knowledge'. See his *The Myth of the Framework.*

[3] J. N. Mohanty, *Reason and Tradition in Indian Thought,* 1992.

[4] Hajime Nakamura, *A Comparative History of Ideas,* 1992.

phers. I shall come back to this point later in the context of Indian philosophical tradition and *other* traditions embedded in local cultures.

In historiography a distinction is drawn between intellectual history and sociological history. The first is internal and the second is external history. General historical account involves "rational reconstruction" of how some events or ideas have grown. But any rational reconstruction requires a normative standard. Indeed the history of science and knowledge is richer than its rational reconstruction. There are always some empirical, sociological or psychological theories to explain the residual non-rational factors. External history can only give an empirical account of the non-rational factors of growth of knowledge, while internal or rational history alone can provide a norm to do so; it tries to explain the growth of disembodied knowledge or concepts. The sociological /empirical factors are necessary components of cognitive idea/science but do not constitute its rationality. Rational history of knowledge and science is selective; it considers the legitimacy of the logical strength of ideas, theories or conceptual frameworks. *How*, w*hen, where* or by *whom* the beliefs are held is of secondary importance to an intellectual history of ideas.

*Time and place* are the two general dimensions of historical research. Periodization – dividing time into periods such as ancient, medieval and modern - is an age-old convention, so is the dichotomy of the East and the West. Moreover religion and philosophy/science in contemporary vocabulary are very sharply demarcated. For a comprehensive history of ideas however, these barriers seem arbitrary. In a study of concepts and theories one should cut across such traditional divisions, for it is possible that an idea or attitude held in occidental philosophy at a certain period finds a counterpart in an oriental philosophical thought, some time even in a religious tradition. Indeed certain intellectual problems are characteristics of certain stages in the history of culture. In different areas of the world similar (if not exactly the same) problems and ideas emerged at certain stages of cultural developments.

Hajime Nakamura[5] suggests that instead of speaking about "East" and "West" types of traditions we should speak of 'cultural areas,' in which some culturally advanced people have established its own cultural tradition in its own indigenous way. For example India, China and Japan are all "Eastern" traditions but despite some partial points of similarity Japanese culture is radically different from that of India. West again is a heterogeneous amalgam of cultures. Greco-Judaic thought is different from say, Anglo-Saxon culture. It is to be noted that 'culture' here includes intellectual elements as well. The degree of cultural developments and diversity again, is different in different occidental traditions. The advancement, say of natural and mathematical sciences, in France and Germany is at a different level from that of many

[5] In his *A Comparative History of Ideas.*

other countries. Logic and Mathematical thought again flourished in Poland and some other places more than many others.

The recognition of cultural factors in moulding scientific and philosophical ideas seems to lead to growing erosion in the sanctity of philosophical and scientific ideas and logical adjudication of ideas and theories seem to be meaningless to several philosophers besides a sociological analysis. Nakamura very rightly contested this standpoint and holds that each study by a specialist, after contextual treatment at some point must be placed in a comprehensive framework to clarify the significance of the total subject matter.

The view that comparative study of ideas and theories belonging to some extent .For each and every element of a conceptual system cannot be entirely compared to that of another system. But *some* elements of one can be fruitfully weighed against *similar* elements in another idea or a theory belonging to a different culture or philosophy. One such possible move would be to concentrate on a "problem" approach rather than an "ism" approach, as Nakamura would put it. I would rather side with a slightly different expression of Popper that problems are more vital than theories, the latter only attempts or conjectures to solve some problems, theoretical as well as practical. In each stage of intellectual development, similar problems arise and responses or ways to solve them can be concentrated upon. New problems and ideas emerge from actual confrontation with problem situations and their tentative solution.

If we rummage through history of science, it would be evident that many scientific theories have actually originated from myths, fables and even from age-old superstitions.[6] The seed of the heliocentric theory was embedded in the ancient sun-worshipper's faith. It is interesting to note that in eastern religion and faith, sun occupied a central and reverential position long before the Copernican theory saw the light of day.

Many apparently unfounded legends gave birth to highly acceptable scientific theories in a later stage of intellectual development. Some cultural areas (societies or nations) again can lag behind some others at the same period. It is well known now that science was not unknown to the ancients. Old civilizations such as Greco-Roman or Indo-Chinese-Japanese have definitely shown rudiments of science with full-fledged theories although not exactly in the same terms as found in today's West-dominated scientific culture. Modern science involves a lot of hypothesis, imagination and extra logical factors. In fact modern sociologists of science tend to strip scientific enterprise of all acclaimed rationality and objectivity which are ascribed to it. Science in this view is nothing sacrosanct purported to aim at truth and ra-

---

[6] Alexander Koyre, *From Closed World to the Infinite Universe* contains many such stories.

tionality. It is a social culture just like any other human endeavour. We shall come back to the issue in connection with modern sociologists of science.

As pointed out before, the most undesirable and misleading consequence of rigid periodic categorization is that it tends to attach some stigma with the categories. "Medieval" thought, for example seems to reek of irrationality and dogma, our ordinary use of language indicates this bias. "Ancient" seems to mean antiquated, defunct or at least similar anti-progressive epithets. "Modern" on the other hand, sounds progressive and is usually associated with all that is rational/logical and scientific; these kinds of associated meanings send wrong signals. If one studies the problem-areas and theories of ancient thought-systems a substantial amount of it can be meaningfully reinterpreted in the light of ideas developed in other stages of scientific thought. What goes in the name of medieval philosophy again contains many impeccable arguments and vast amount of problems and theories worth a thorough re-examination. In fact recent spurt of research in these areas in the west confirms that point. Sophisticated comparative descriptions of various philosophical systems were not unknown in India at that time. *Sarvadarshana-samgraha* (literally meaning collection of all philosophical thoughts) by Madhava (c1350) is one such masterpiece .It is true that before the advent of modern science religious supremacy reigned over free individualistic thought. In pre- modern era institutional authority dominated cultural life both in eastern and western society. The negative dictum of "Thou shall not" pre-dominated over "Thou shall" in social life. But social discouragement notwithstanding, free scientific thought was not completely weeded out. Otherwise great scientific theories would not have been possible. From Socrates to Galileo it is the same story of suppression and persecution but still the great ideas survived and were improved upon by posterity. We can remember the story of persecution of our own Kshana in ancient India for her astronomical observations and mathematical calculations but still the scientific ideas survived and outlived social dogma of the time.

Again if modernity implies emphasis on perception and reason as source of knowledge then it has been a part of Indian tradition of philosophy. If elaborate systems of grammar and language as well as theory of proposition and syntactical rules are any criterion, the same tradition exhibits sophistication of a high order[7]. Highly developed monism and realism are also to be found there.[8] Moreover otherworldliness and faith in supra natural/empirical existence were contested, like modern theories, by some "atheist" systems of

---

[7] B. K. Matilal, *The Character of Logic in India,* 1999, contains a good account of such a claim.

[8] P. K. Mukhopadhyay, *Indian Realism : A Rigorous Descriptive Metaphysics,* 1984, as the title shows is an attempt to reinterpret Indian concepts of realism as well as monism.

traditional Indian thought. Astronomy and mathematical sciences attained a considerable height in ancient India. Along with it aesthetics and logic (including the notion of class and the distinction of language and meta-language) were quite sophisticated. A comprehensive investigation of the great Asian traditions of China, Japan and India may indicate that the traditions were not unprepared for the advent of modern age. The same is true about any other ancient tradition, eastern as well as western

Two kinds of knowledge can be distinguished: universal (Reason, Morality - monolithic Truth) and particular (traditional, pluralistic, multi-culture/morality, truth and reason, immanent in the various traditions). The first one is the norm of objective rational knowledge. Like truth, moral standard is also independent of the frameworks. Knowledge must be justified on good ground. Science is the paradigm of rationality. Reason is objective. Truth is one. Standard of validity is universal. Knowledge *transcends* the framework of culture, religion, language and other social conditions.

The universalist model of knowledge and rationality faces many internal problems in its pristine form as presented by the received view held by Carnap and many other philosophers and scientists in the earlier half of the twentieth century. The dissatisfaction found many responses in different directions. One direction took the form of rejection of the justificationist model of scientific knowledge, which amounts more or less to say that scientific knowledge is based on empirical evidences; therefore it can never be justified. In fact only trivial truth (like 'p is p') is verifiable/justifiable and it is apparent that these types of propositions are cognitively empty. Propositions that are cognitively rich however can *never* be justifiable. On the contrary more informative content makes them cognitively more rich and 'interesting' but at the same time more vulnerable to be falsified. As is well known Karl Popper holds on to this type of critique of the justificationist model of scientific knowledge and at the same time argue for a non-justificationist but rational model of truth and knowledge.

The other critical move in this respect is the total rejection of *all* kinds of attempts that distinguish between rational and irrational or true and false. Only a sociological analysis of scientific theories is carried out. In fact the distinction between scientific and non-scientific is blurred by these attempts. It is made also to argue that there is no such distinction. This move is generically known as sociology of knowledge as it attempts to give only a sociological analysis of scientific ideas and theories. It tries to reduce knowledge to only a kind of social culture no more than any other culture. Before going into an examination of the postmodernist critique of scientific knowledge, let us first examine some of the salient features of what is generally known as sociology of knowledge.

Kuhn in the 'sixties and the sociologists of knowledge from the 'seventies emphasised the cultural nature of science and social determination of knowledge. According to this view, science is nothing but a culture. There is no universal standard of truth and rationality. Anthropological research leads to a kind of relativism that espouses different standards and norms for accepting a belief. Rival paradigms determine different kinds of belief-systems in which a statement/belief would be rational if it is coherent with other beliefs of the system (example: "Black people are inferior", coheres with other beliefs in a racist society. "Lower caste people cannot marry a Brahmin" similarly would be rational in a caste-ridden Indian milieu). Rival paradigms or scientific theories similarly give rise to propositions, which follow from the paradigms (example: Copernican/Ptolemy or Euclidean/non-Euclidean geometry). The notion of truth and rationality thus is not absolute—but relative to a particular standard. No knowledge is thus universal; all knowledge is standard-specific, relevant only in a particular context. There is no universal norm to judge the absolute truth or rationality of knowledge. All epistemic standards are relative and determined by social conditions: one thing we should remember here: nobody anymore would subscribe to the naïve view that knowledge is absolute but that does not preclude the possibility that it is objective.

The proponents of relativism of various sorts put before us standards of mutual understanding and possibility of knowledge unrealistically high (not very much unlike the classical sceptics). And when we fail to meet those standards, they claim that understanding of those standards is impossible and therefore no cognitive status can be conferred upon *any one* of them. Karl Popper[9] is the greatest champion to explode what he calls *the myth of the framework*. While he admits that there are different cultures including cognitive traditions and there is great deal of social factors influencing them, he does not think these factors are relevant for the validity of those traditions. It is also granted that it is very *difficult* to overcome the influence of the framework but it is not *impossible* nevertheless, to overcome that difficulty by rational /critical discussion.

Unlike the relativists Popper regards confrontation of different culture as contributing to progress of knowledge resulting from what he calls "culture clash". History of western culture, science and civilisation can authenticate this point.[10] Although he is an ardent admirer of tradition at the same time he is 'an almost orthodox admirer of unorthodoxy' and regards orthodoxy as the death of knowledge as mentioned before, for 'the growth of knowledge depends entirely on the existence of disagreement'. Indeed disagreement may lead to strife and even to violence, which is undesirable but it may also lead to mutual criticism and discussion, which are necessary ingredients of progress of knowledge. He cites the history of development of western science

---

[9] Karl Popper, *The Myth of the Framework.*

[10] Ibid., Section III.

and philosophical ideas as a result of clash and confrontation of different frameworks. The strife of Greco-Roman civilization as well as their clashes with the Jewish, and other Middle Eastern civilizations brought forth advancement in knowledge. Progress in knowledge is also indirect fallout of clashes with Germanic and Islamic invasions.

It is not very clear what Popper means by 'culture clash'. When it is conferred a positive role in the rise of Greek science, it seems he means that interaction with different traditions contribute to a better understanding of the world and its various problems. It also includes conceptually more rich ideas about nature, society and political and moral issues. Of course the most fundamental among these ideas is the idea of rationality. The tradition of rational and critical discussion shows the advent of democracy and the spirit of scientific tradition. Popper refers to the rise of Greek democracy as well as scientific ideas. We can similarly refer to the Indian tradition of *vada* (*hypothesis*), *and vitanda* (*criticism/debate* ), mentioned in the *Nyayasutras* and also the *nigraha-sthana* (defeating situation) in manuals. These methods determine win or defeat for a debating position in early Christian era. This tradition led to a gradual switch over to a rational and argument-based approach to solve intellectual and practical problems. Popper's crusade against any form of relativism is well known, although he regards all knowledge as fallible and science only as a progress *towards* truth which may not be attained forever. However Popper thinks rationality as the most fundamental idea related to our understanding of the world and thus gradually to the rise of science.

'In its application to the problem of understanding our world, and thus to the rise of science, rationality has components which are of about equal importance'.[11] One of these two components in his view is poetic inventiveness or the invention of stories, which explains the world. It is rightly pointed out that in all cultures whenever there is some thing that needs explanation, people make conjectures, often wild conjectures that created different kinds of myths about unknown powers to understand the world and the mysteries of human life which they cannot fully fathom.

It is true that this component is perhaps as old as human language itself and seems to be universally present in all traditions, 'all tribes, all peoples have such explanatory stories, often in the form of fairy tales. It seems that the invention of explanations and explanatory stories is one of the basic functions of the human language'.[12] The second component, which he thinks is comparatively modern, is found with the ancient Greeks, particularly with Anaximander, the second Ionian philosopher. 'It is the invention of criticism,

[11] Ibid., Section V.

[12] Ibid., p. 29.

of the critical discussion of the various explanatory myths, with the aim of consciously improving upon them'.[13]

The main Greek example Popper gives of explanatory mythmaking on an elaborate scale is that of Hesiod's *Theogony,* which is a wild story of the origin, the deeds and the misdeeds of the Greek gods. In Indian mythologies we have comparable stories of relating natural phenomena with divine wrath or pleasure and even hierarchy among Gods along with their special characteristics to explain problems about human life and natural phenomena that we do not understand. Anaximander's break away from Thales' myth (that the earth needs some support for its stability) by maintaining that nothing is holding the earth - gradually made the bold idea of Copernicus possible and it even contains the seed of the idea of Newtonian forces. Examples abound in both Indian and other mythologies to indicate that traditional perception often led to some more reasonable explanation of the world and more critical understanding of the same.

Popper regards Anaximander's example of breaking away from the tradition of a support for the earth as a bold step to a more significant problem of the issue of the support and stability of earth. This initiates a new tradition – the critical tradition. There may be difference of opinion among historians and philosophers of science whether it can be called a *critical tradition* but one thing is at least clear that *a critical tradition (read scientific tradition) is constituted by adopting a method of criticising* a received story or given explanation and then going forward to a new, improved imaginative story which explains better. But by all means it is subject to further scrutiny or may be replaced by a better story or theory if you like. Undeniably this is the scientific method. Indeed lots of non-rational elements are there along with rational ideas in our attempts to understand the problems of the world. The sociologists of knowledge are right about it; they are also right in pointing out that it is a very difficult task to disentangle the non-rational from the rational elements of our understanding of the world. But from that difficulty we cannot conclude that the task is impossible.

Rational approach does not assume either infallibility or demonstrability of any idea or a theory .It seems that the positive aspect of what Popper calls "culture clash" is the growing recognition of other standpoints and perspectives. These may first seem to be very alien but a rational adjudication and understanding may reveal their worth. But mere recognition of alternative viewpoints does not also guarantee that these viewpoints are of the same value. Nor does it imply that none of them are more than reflective of a certain social/cultural condition. Translation or interpretation of other traditions is a great obstacle no doubt, and various cultural, political factors are the main hindrances on the way. Historical events and contingent conditions

---

[13] Ibid., p. 29.

such as colonization and imperialism are principal factors that made it difficult for the victors or the colonial powers to appreciate the meaning and value of the culture of the vanquished.

It is well recognised now that the indoctrination with western ideas is a major hindrance that inhibits understanding of other traditions, says eastern traditions of understanding and attempts to solve similar problems. By this statement it is not intended to mean that *all* eastern ideas are of equal footing with comparable Western ideas or that either of them is better or worse than the other. Each and every idea or theory is to be understood with reference to the context it is made and weighed individually for its value. For example Ptolemy's theory was consistent with the physical theory of the day and empirical observations; Copernican theory was eventually vindicated against it by the adequate physics of Newton. In this connection we can refer to Popper's concept of 'culture clash' and the possibility of inter-translatability of language (contra Quine) in principle notwithstanding the fact that it is sometimes a very formidable task, given the actual cultural difference.

Moreover culture clash would lose some of its importance if 'one of the clashing cultures regards itself as universally superior and even more so if it is so regarded by the other'.[14] The reason for such loss is clear, if one of the cultural rival becomes convinced of its inferiority, then the critical attitude is lost and hence no good consequence comes out of such intellectual confrontation which requires both the parties to be of similar strength. Moreover 'whenever there was a major obstacle to overcome, it was as a rule the result of indoctrination with Western ideas'.[15] These two citations from Popper indicate his closeness to the sociological theory of knowledge in recognising the active role of social factors in theory formation. His theory is often confused as a logical theory that does not take the socio-cultural factors into account. The only great difference lies in the recognition of these factors having relevance *only* with reference to the origin and nature of the theory or idea about the world and *not* about the rational validation of those ideas. Scientific tradition is different from all other traditions in offering a methodology to test the value of its ideas, which may very well turn out to be false/invalid. It is by no means different from other traditions in leading to the realm of infallible or at least demonstrable knowledge about any empirical matter except trivial truth.

The moot point is not to equate reason with modernity alone or else the only alternative is not to banish reason from the realm of science because it is also a socially conditioned culture. On the other hand we may also consider some traditional idea and mode of reason to have the potential of good logical power. It is not however maintained here that all systems of ideas are of

[14] Ibid., Section, IX.
[15] Ibid., Section, IX.

equal strength as these are products of some social condition or other as the sociologists of knowledge presume. Rationality and objectivity ideal of science and knowledge has become an object of indictment by various critiques of the Enlightenment ideal which reject the rationalist model of knowledge and embrace a model constituted only of particular extra-logical norms and sociological standards.

Post Enlightenment liberalism has a big problem to face: how can Enlightenment universalism be abandoned and yet avoid a collapse into radical particularism which leaves liberalism as no more than one form of life amongst others with no privileged claim upon human reason? John Rawls and Richard Rorty have abandoned the pursuit of rational foundation for articulations and celebration of the confessedly particular ideas of a particular tradition, of 'certain ideas ... latent in the public political culture of a democratic society'. Widespread reaction against the Enlightenment ideal, age of reason and liberal universalism, and a growing awareness of the tensions inscribed within multicultural societies give rise to a gradual breakaway from this ideal of truth and rationality. Sociologists of knowledge and science in recent times as mentioned before uphold that truth and rationality are *not* universal or one, they vary from one framework to another. Science seems to be the typical rational model of knowledge and universal truth which is generally the target of such attack. Thus Rorty's open and unapologetic grasping of the particularist horn of this dilemma has left him vulnerable to the charge that he ultimately lacks the courage of his lack of conviction, and is compelled covertly to invoke universal claims about human nature in support of his liberal allegiance. The moot point however is to override local beliefs and commitments in the name of logic and universal ethic and (particular) concept of respect, i.e., refrain from imposing principles on others that do not follow from their own belief-systems.

Notice that this debate is *within* the framework of a typical western scientific model. It does not include non-western or/and non-modern knowledge-systems. If we may call this latter kind *traditional* knowledge, it may include both non-western types of knowledge and non-modern model which may still be prevalent in society where many people still live in traditional way. Ancient and vast countries like India, despite its enormous progress in science and technology still have a substantial population (like marginal farmers and forest-dwellers) who rely more on traditional wisdom than modern western science. Their standards, axioms and theories are different from the standardized universal knowledge like western science. Their standards, axioms and theories are different from that of other traditions. The reason for regarding them as alternative knowledge system is that the hypotheses are consistent with the axioms and other hypotheses of the system; In traditional Indian systems of knowledge (I am sure it is equally true about other ancient cultures such as Chinese, Egyptian, Mexican etc.), there are full-fledged theories of the universe, astronomy, mathematics, medicine, agriculture and so on

## *The Case of India*

Basically there are two strands of knowledge: (1) The Brahminical tradition of theoretical knowledge linked up with some philosophical system or other. (2) Traditional local belief-systems that aim to solve practical problems rather than justifying a philosophical /scientific theory.

The aim of any adequate theory of knowledge is to solve some problems and seek explanation for events unknown. All cognitive projects consist in attempts to answer some questions like *how do I know that this is so and so?* Or *why this is so and so?* Or else *how to solve this problem?*If problem-solving is a feature of all epistemic projects it may either be purely theoretical, that is relating a theory or conjecture to a larger system of ideas or else solving some immediate problem, cognitive or merely practical (which does not necessarily exclude the possibility of being upgraded to a more serious idea). In that case many forms of traditional knowledge, no matter whether it is a part of an ontological theory or handed down by a local system of knowledge to the posterity - fulfils that requirement and cannot be denied an epistemic status.

With this consideration in mind it is not so clear that *only* universalist thesis is correct and the *only* form of valid knowledge since no non-trivial proposition can claim to be fully justifiable. If this point is granted then it is not so easy to dismiss various forms of local/traditional knowledge by relegating it to a non-epistemic status, for after all these forms have reasonably handled various problems for ages and helped people of these traditions. I consider two vital problems, perpetuation of life and conservation of nature were handled if not solved quite successfully in India by both philosophical and local/traditional methods. Feyerabend's apprehension is well founded that many problems of life were handled quite successfully and reasonably well by non-modern/non-western systems of knowledge. But the important point is that problem solving is *not* the only function of a scientific system. It is often overlooked by Feyerabend and other champions of traditional knowledge that prediction of possible state of affairs is equally important norm for a scientific theory. It is true that the traditional systems incur as little interference with individual freedom as possible unlike modern scientific approaches of various kinds. As is well-known he cited the case of perpetuation of life by referring to different traditional medical systems that respect autonomy of the individual, stress preventive measures and medicines rather than complicated diagnostic processes and treatments that violate individual dignity of body and mind.

I shall however emphasise the other area, namely conservation of nature that has become a glaring example of destruction by various development projects which are now essential indicators of modern science and technology. Most critiques of science and modernity point towards traditional methods that were less drastic and thus more congenial for conservation processes.

It is possible to formulate an account of rational ecology which is in tune with the basic tenet of individual freedom in general and to argue that all ecological theories are more or less anthropocentric and thus cannot provide a good reason why we should preserve nature, for all kinds of development-related technological projects need some degree of natural devastation. So we require some rational principles that are *not* anthropocentric and include nature in general under the purview of individual freedom .It seems that the critical rationalist position which espouses criticism and individual freedom is compatible with this ecological position which respects individual freedom and at the same time encourages development and progress with least interference with nature. The principles of rational ecology do not require all traditions to be equal but at the same time discourage mindless devastation no matter who are the main beneficiaries. Before going into that, let us have a quick look at the epistemic traditions of India and how local knowledge played an important role in various ecological problem solving.

(1). Philosophical knowledge in ancient Indian literature is found in the different Vedas, Samhita (the physical universe, astronomy, medical science, chemistry, logic and grammar, physical and mental health (yoga), etc.

(2). Local knowledge of flora and fauna handed down from the ancestors, myths and legends are not necessarily related to a theoretical framework, but helpful guide for perpetuation of life and conservation of nature. Indeed, traditional livelihood relies heavily upon nature for its sustenance, but the need of such people is not much. Unlike the traditional way, modern commercial exploitation of nature can expedite denudation in a degree of destruction disproportionate in comparison with the degree of regeneration and growth. Some local beliefs/superstitions also helped conservation if we compare it with objective knowledge of modern science and technology. Examples abound in literature of traditional forms of knowledge aiding both perpetuation of human life and conservation of natural resources.

In pre-industrial time man was closer to nature in the sense of accepting it as such and interacting in a more direct way with no aggressive attitude towards it. In other words, there was a give and take relation with nature without more demands on it than the basic necessities. This attitude was prevalent in traditional India which was predominantly an agricultural society based on village communities and various mores rooted firmly on traditional values. The strong bonds of family and clan-sentiment led to respect nature in its diverse forms: rivers, mountains, plant and animals with reverence and love.

This attitude is regarded by intellectuals as animistic, i.e., attributing spirit or mind to inanimate objects. Instances of such respect for nature and other creatures are plenty, for example it was common practice in India not to milk cows for quite a while after calves are born, or not to tear leaves, pluck flowers or fruits from trees at night when they are supposed to 'sleep'.

All these attitudes are changed by the so-called epistemological conquest of nature by science, which claimed to "know" and dominate nature so that the latter can serve human beings, by its enormous resource potentials. Even the problem of conservation of wild life was handled much more successfully here than it was done in the west. Given the long history of habitation in the Indian sub-continent the variety and number of wild animals are still reasonable despite mindless poaching and destruction for either as a game or for the motive of monetary gain .In comparison to those wolves are totally extinct in mid-nineteenth century England. In America wild buffaloes now belong to a rare species.[16] R. Sukumar argues that despite conflicts and confrontations between humans and wild life people in the sub- continent have dealt with wild animals such as "rogue" elephants and man-eater tigers and other ferocious animals through the ages in a reasonably effective but tolerant way. Otherwise these animals would not simply have survived to this day. Especially if we consider the long history of settlement in our country and the large human population directly and indirectly dependent on forest for sustenance, it is evident that the problem of wild life has been satisfactorily tackled. Elsewhere people have solved their problem by simply wiping out an offending species.

As said before during the nineteenth century, wolf a predator on sheep in Great Britain was completely exterminated. In the past religious or other taboos helped conservation in a more tolerant way than now. But we know that large mammals like elephants and tigers play an important role in an ecosystem by virtue of their large biomass or position at the summit of an intricate food web. Disappearance or depletion of such key species could lead to the disruption of ecosystem functioning, perhaps leading to extinction of other species as well.

Thus the symbiotic relation in which nature and man lived together for millenniums have been snapped by the onslaught of western science and colonialism in less than three hundred years. In this short period nature has been destroyed, depleted and exploited in an unprecedented degree which our five thousand year old civilization never witnessed before. The worst part of such conquest is that only a minority privileged class is benefited by it and

[16] R. Sukumar,'Wildlife – Human Conflict in India', R. C. Guha, ed. *Social Ecology*.

the fruits of such exploitation never reach those who live nearest to nature and have drawn sustenance from nature for thousands of years. [17]

Apart from the specific account of exploitation of nature by the colonial power and now the various governmental and non-governmental agencies, this kind of exploitation of nature in the name of industrial development, many ecologists think, is not only wrong but also myopic about its disastrous effect and future consequences. An ecologically rational man-nature system is one in which human and natural components stand in a symbiotic relationship. Besides this the rationale of the basic norm of "live and let live" corresponds to the norms of traditional philosophy to a norm of 'maximum fulfilment of life potentials' as J.S. Dryzeck[18] puts it. This norm implies norms of diversity, complexity (between man and nature thus of structure and function) and of maximum symbiosis as mentioned before that is arrangement of life forms and life styles with minimum negative interference with each other.

Another point we have to note is that the so called natural classification also is socially determined, why for example, water hyacinth is to be destroyed while corn and sugarcane protected? The reason is obvious that whatever is suitable for human consumption and at least enjoyment (a rose or sandalwood) is to be preserved and the rest is to be destroyed or at least minimized to make room for human need fulfilment. The broken balance between man and nature thus creates an uneven and paradoxical picture.[19] Mukherjee maintains in similar vein with the modern trend of rational ecological principles that, 'each community cannot appropriate more than its due place in the general ordering of life from which nothing can be obtained without influencing everything else. Working symbiotically, they represent interwoven threads of a complex web of life'. No threads can be snapped or removed "without the whole garments of the life and nature and human society being disfigured'.[20] The ecological knowledge and awareness of the new millennium echoes the old fashioned metaphors used by him almost seventy years ago to describe a hypothetical state of affairs which we actually face today.

The other reason to have discontent with universal scientific knowledge is that it is tyrannical in a certain sense. The sacrosanct model of scientific

---

[17] R. C. Guha, ed. *Social Ecology* and *The Unquiet Woods* contain various accounts of such exploitation and resistance movements in India.

[18] J. S. Dryzeck, *Rational Ecology*.

[19] Radhakamal Mukherjee, a pioneer in ecological research in India mentioned such examples of imbalance man brought into nature. He says: 'The plants which he (man) does not tolerate are called weeds', see his 'An Ecological Approach to Sociology :Eco-contributions to Sociology', *The Sociological Review,* vol. xxii, No. 4, Oct. 1930, reprinted in *Social Ecology*, ed., Ram Chandra Guha.

[20] Radhakamal Mukherjee, ibid ., p. 26.

enterprise makes modern science as well as its practitioners hold a high status in the role of a neutral arbiter of truth and rationality, disregarding other alternative epistemological traditions. The disillusion with modern scientific model led some philosophers even to doubt the so-called ideals of knowledge for various contingent reasons which I have discussed elsewhere[21]. Indeed the problems related to the domineering role of science and technology cannot be overestimated. Feyerabends' fiery diatribe against science and scientists has many substantive points. It is time now to be aware that modern western science is engulfing other traditions. Even concepts and language are entrenched by it. Why for example, western culture is mainstream and other cultures 'alternative 'cultures? Surely eastern civilisation has at least some claim to be temporally prior. The division into 'first world', 'third world' or 'developed' and 'developing' (a patronising change from 'underdeveloped') again is arbitrary and questionable.

The favourite example of the tyrannical nature of scientific knowledge is modern medical science where more transparency is desirable. It is also true that scientific research is guided more by political factors than human cause, but all these considerations at most indicate, that actual scientific enterprise *has been* often tyrannical, and *not* that there is something basically anti-democratic. I have also argued elsewhere that no protective measures (monitoring, supervising etc., as suggested by Feyerabend) can help, for science itself is a free enterprise and scientists also are free individuals. So if there is any tyrannical element in any enterprise then it must be dealt with from *within* that very culture by methodological devices and transparency. Scientific research provides with lots of checks and balances from within the tradition. Therefore the actual state of scientific research cannot provide us with a reason to call it tyrannical .As said before modern science has been misused, and often assumed a domineering role but despite all these deficiencies it has been so far the most reliable cognitive tradition, for its methodological norms.

One point however has to be taken seriously regarding the value of other cognitive traditions which recognise many basic values so far ignored by the universalistic attitude of modern science in sweeping all problems (and their solutions) under a standardised method and analysis. Many problems of a specific nature can be solved by some age-old ways more successfully than

---

[21] 'Tyranny in Disguise" - Expert Knowledge and Human Interest, *Facets of Humanism, ed.* Patnaik et al, New Delhi, 1999. Radhakamal Mukherjee, a pioneer in ecological research in India mentioned such examples of imbalance man brought into nature. He says: 'The plants which he (man) does not tolerate are called weeds', 'An Ecological Approach to Sociology: Eco-contributions to Sociology', *The Sociological Review*, vol. xxii, No. 4, Oct. 1930, reprinted in *Social Ecology*, ed., Ram Chandra Guha. Radhakamal Mukherjee, ibid., p. 26.

the standard methods of modern science. Feyerabend also pointed out a significant feature of most traditional methods which interferes in least degree with the area of investigation, be it human body (in case of medical treatment, or diagnosis) or other experiments as it is done in modern science. As mentioned before, most traditional methods are tolerant and least interfering in dealing with their subject matter or problem area

In traditional Indian agricultural system after each cultivation, for example, lands were not tilled for a long period, sometimes as long as a whole year. Unlike this custom, in modern scientific agriculture multi-crop system gradually robs the natural fertility of the soil leading to various problems including excessive use of chemical fertilisers which again causes or at least can cause harm to human and animal life. Similarly cows were *not* milked for at least some months after a calf is born, unlike modern scientific dairy system where this kind of consideration is absent to make room for large-scale production of milk and other dairy and poultry produces. Many herbal and other natural ingredients also were used as aphrodisiac as well as contraceptive without the hazard of harmful side effects. These are but a very few examples to show only the kind of way traditional knowledge operates. Some of these knowledge are merely part of the "little tradition" handed down from customs and local knowledge, but there are many others which are not merely part of the tradition but can be related to some philosophical or ontological theories. The latter type of knowledge can therefore be called scientific because it is consistent with and follows from a general view of the universe.

Local knowledge however can be either related to a philosophical theory. Or else it can degenerate into nothing but a mere local custom which still is not connected with other ideas or theories. But the possibility can never be ruled out that an idea can be in principle related with some theory or other. The ancient way of sun worshipping for example, led eventually to the heliocentric theory of the universe .The sacralisation of nature in similar way is a primitive way of respecting and understanding ecological problems and conservation principles.

As mentioned earlier in the discussion, J.N Mohanty's analysis[22] of the Indian scenario of modernity and tradition is worth mentioning in this connection. He rightly thinks that modernity and tradition are perfectly blended in Indian political thought and not so much in academic philosophy as mentioned before. This has been possible by linking theory with practice. The Gandhian model of combining ideology with practice is a good example. Mohanty thinks that the Gandhian ideas of decentralised village-centred economy and rejection of party politics in favour of direct and village centred

---

[22] J. N. Mohanty, 'Indian Thought: Between Tradition and Modernity', *Explorations in Philosophy*, chapter four.

politics has worked towards 'modernisation of Indian tradition on the one hand and on the other, traditionalisation of western modernity for India'. The modernisation is achieved by freeing the traditional society of its age-old concepts of 'otherworldliness' and fatalistic tendencies, which is mismatched with the modern ethos of work culture and individual enterprise. Traditionalisation of western modernity is achieved on the other hand by 'offering a moral critique of modernity, industrialisation and bureaucracy and advancing a set of norms to guide the political structure of bourgeois modernity'.[23] The idea of *dharma* (duty) plays a pivotal role in traditional Indian ethos although the concept is much maligned in the modern context and strayed far away from its original intention and purpose.

Gandhian notion of combining tradition and modernity is unfortunately perceived by many as irrelevant for current concept of liberal thought that emphasises competition rather than moral dimension as more important element in the advent of modernity in India. However there are still some thinkers who still hold that Gandhian concept can be remodelled in the context of modern India and is a better solution to resolve the riddle of combining tradition with modernity.[24] Rajni Kothari regards Gandhi as an 'alternative to Western post-enlightenment thought and to its related ideology of developmentalism and on the other as representing the moral imperative of treating people as a source in the recovery of a humane order'.[25] Mohanty thinks Kothari's interpretation can be seen as India's 'own version of post-modern and post-secular society, with its own tradition of social pluralism (as distinct from the political pluralism of western democracies) and a new concept of unity and its high tolerance of ambiguity and scepticism'

Nandy however perceives Gandhi as the most "creative critical-traditionalist" who accepts, combines and at the same time critically reviews both. He regards Gandhi's rhetoric and reference to traditional Indian myths as powerful instrument to overcome the dominant Eurocentric concepts in various areas. Some of these concepts are scientism (which has become the focus of criticism in the west much later), historical determinism (strongly critiqued and rejected by Popper) and technicism etc. These concepts attracted attention of many western philosophers. They recognised them as the

---

[23] Ibid., p. 61.

[24] Rajni Kothari and Ashis Nandy are some of these thinkers who are trying to support the basic Gandhian principles in the context of modern problems.

[25] Cited from Mohanty, Ibid and Thomas Pantham, *Political Theories and Social Reconstruction : A Critical Survey of the Iterature on India* . I am indebted to J.N. Mohanty's own views and his reference to Ibid., page 61. Rajni Kothari and Ashis Nandy are some of these thinkers who are trying to support the basic Gandhian principles in the context of modern problems.

root cause of much modern malaise that afflicts us in the present and would be aggravated in the future.

Gandhi rejected the historicist's dogma of historical inevitability of the future not very much unlike the Popperian critique of it without the latter's systematic arguments against the doctrine. Gandhian principle of local /partial reformation rather than wholesale social change also has an implicit cognitive import. It can be phrased as follows: The future is indeterministic and therefore not known or predictable; any holistic approach to social phenomena may have undesirable consequences, therefore small scale social change at least cannot bring forth a great disaster even if it does not bring a great fortune or total change. Therefore small changes, political or economic are preferable to wholesale change like revolution on the social/ political spheres and big capitalism on the economic front. Similar arguments are offered by Popper for his famous theory of "piecemeal engineering" for social change which only is consistent with the rejection of historical determinism and emphasis on human fallibilism. Gandhi's view of local rather than total view of social change however is based on a moral basis of self-sufficiency, self-determination and self-criticism etc.

In an attempt to harmonise tradition and modernity one may hold a position that modern scientific knowledge is a stronger tradition and traditional local knowledge handed down by local practice needs protection. So long the universities were 'the repositories of modern scientific knowledge'. But as Ashish Nandy maintains, modern scientific knowledge-system does not need as much protection as traditional knowledge-system, therefore future universities should act as legitimising authority for different types of these traditions before they are forgotten or at least dropped from practice[26]. I completely agree with this stand and want to emphasise that by such endorsement it is not also claimed that *all* traditional knowledge can stand the test for acceptability. It is only argued that they should be given a fair chance to be tried along with modern science which is already on a much firmer ground .It is very well possible that much of traditional practices would not prove to be of much worth. But at the same time we cannot discount the possibility of retrieving something from the jumble of complicated social practices some of which at least may be counted as reasonable. After all, these traditions for example as in India have a long history of survival despite many unfavourable natural and social factors. Therefore these kinds of local knowledge need protection or at least patient hearing for a fair trial. This requirement does not by any means undermine the value of modern science but only demands a way to keep open the possibility of retrieving something from traditional local knowledge-system.

---

[26] A. Nandy, 'The Role of Future Universities', *Seminar*, June 1996.

## *Social Determination and Social Construction of Knowledge*

Against the Universalist thesis, it is said that science is a kind of culture, so it is determined by social factors. A stronger position is that the facts science examines are not culture-independent, they are theory-specific objects, nothing but constructs of a particular scientific tradition. In recent discussions about anthropological universalism the notion of 'world-openness'is the main focus. Some philosophers think in the line of German philosophical-anthropological tradition that world openness marks a transition from nature to culture .It can be argued first that this concept being ambiguous blurs important distinctions between openness to determination by cultural forces and capacity for self-determination

Secondly, the use of the concept of either variant is compatible with radical cultural relativism, as it appears to allow for the deactivation of any universal biological dispositions[27]. It is claimed by Roughley in the "Afterword", "Human Nature, A conceptual Matrix", that conceptions of the human enable the distinctions of four senses of the expression 'human nature, the pre-givenness of properties and their essentiality for being human is one of the three ways: as conceptually, predicatively or axiologically central'. It is argued that discussions of human nature need to be aware of the multiple references of the term and to mark clearly which of the senses they are discussing. Particularly when the author wishes to argue that there are specific relationships between structuring role in theory construction in philosophical ethics, aesthetics and social sciences.

It is rightly claimed also that these roles are often unacknowledged or implicit. It can also be reasonably said that enlightenment ideal for reason and morality (science and the concept of good) stood for western expansion and colonialism. In a sense enlightenment becomes another name for empire or domination of western power over others. Roughley rightly adds, 'exoticization of the non-western world is consequence of a historical force – the postcolonial celebration of the other – the mythic, the primitive (the ethnic), the non-western is deeply implicated in a system of global segregation'. Without denying the importance of the actual impact and historical significance of the point made here it can still be argued that these historical consequences thereof does not show fundamental flaw of the rationality ideal but *only* its actual misapplication. The ideal is a general schema for what should be considered as good reason or good action, without making any further claim.

---

[27] Neil Roughley (edited), *Being Humans*: *Anthropological Universalism and particularism in Transdisciplinary Perspectives*, 2000.

One can make a distinction 'between dialogue within a knowledge-community and *between* knowledge-communities'.[28] Zhao Tingyang argues that understanding *other* communities does not entail acceptance of them. Western epistemological community regards 'others' only as interesting and exotic, but do not accept them as equal alternatives. . Western universalism assumes that other systems of knowledge work in the same way. Tingyang claims that unlike western reason-based approach, Chinese philosophy emphasises appeal to the heart. Practical negotiation, development of a reciprocal idea rather than logical argument characterises the latter. 'In a sense Universalism is Fundamentalism, a negation of the other', in this view.

One of the successful ways of understanding the other in the last decades is an anthropological epistemology, the contextualized local knowledge by the "thick description", first recommended by Ryle as a philosophical task, and of the other culture as Geertz encourages. The epistemological turn from the universal knowledge to the local knowledge can provide us with more information of the other, but the problem of understanding the other still remains even in the 'ideal speech situation'. The so-called pluralism thus is only 'deconstruction' of universalism - a diluted version aiming at politically correct attitude to others. Tingyang thinks that an interactive reciprocal approach (like the Chinese way) aims at an appeal not to logic and reason but to 'heart', happiness rather than knowledge, is the more comprehensive way to understand epistemology. This sounds very similar to Feyrabendian clamour for ordinary reason and human happiness rather than scientific rationality and logic as the real goal of scientific epistemology. One point must be noted here, granting that it is difficult to understand other culture-based viewpoints we may also grant that understanding does not guarantee acceptance as Tingyang argues. But we may notice that whenever there was a major obstacle to overcome, it was generally a result of indoctrination with Western ideas. Or it comes from at least dogmatic acceptance of some uncritical scientism amounting to nothing more than Western ideology. This is no less a grave danger to critical thought (as Popper would have put it) than succumbing to age-old superstitions nurtured by orthodox traditions.

No one can deny the social factors for an idea or theory. As no theory can originate in a vacuum, social factors are always important necessary conditions for a theory/knowledge. An idea needs nurturing in a congenial atmosphere; otherwise it cannot thrive into a full-fledged theory/knowledge. Example: In ancient Indian mythologies and epics we find reference to objects like rocket/missile, flying saucer, aeroplane, amphibious vehicles, etc. But due to many contingent historical conditions like internal strife, invasions

[28] Zhao Tingyang, 'Understanding and Acceptance', paper presented in Universal and Local knowledge seminar, Sorbonne, October 2001.

from outside over the ages, colonialism - these ideas are *not* first developed by the Indians but invented over the years by modern western science.

But social conditions are *only* necessary conditions for knowledge or rather for its origin. Knowledge/theory needs to be justified or legitimised by some logic or at least heuristics. That process transcends social/local factors. Such rationality is an accepted norm in a civil society. Its denial also has a dangerous moral corollary - any action by any social/cultural group can gain an approval by such sociological/local standard of rationality and good. The world has witnessed much destruction of humanity in the name of ideology during the last century and also at the present. No sensible epistemological norm can dismiss it as only an alternative system of knowledge and ideas. Moreover, although there can be multicultural standards belonging to different traditions, the standards can be adjudicated by some reasonable norms of consensus.

The purpose of philosophical investigation is to help people make evaluation of their beliefs and actions using the most objective and least personal or idiosyncratic criteria that we possess. It can be done by articulating fundamental standards to utmost scrutiny and careful analysis. "Our understanding of basic epistemic and moral principles may be imperfect, but this fact provides no reason for describing every disagreement as merely a matter of personal history, societal position or cognitive taste".[29] For India both the neo-Gandhian and the 'left' inclined post modernist critique of science and rationality has taken a peculiar turn faced with a consequence which is least desirable even for them. It denies the universality and objectivity of modern science, thus legitimising all local (traditional) knowledge to be at par with scientific knowledge.

This kind of epistemological populism does not help those in the non - Western societies. On the contrary those who are in these countries need a strong and substantive equality in knowledge to combat various social inequalities prevalent there. We in India should realize the potential of the Enlightenment ideal which can defeat the various kinds of fundamentalism that is threatening the porous firmament of secularism in India. Both the neo Gandhian and the left inclined intellectuals share the post modernist and post colonial suspicion about reason and the Enlightenment, the only difference is the anti-essentialist stance of the latter. They agree on the cultural embeddedness of all ways of knowing which reduces modern science to a mere 'ethno-science' of the west. The whole project of 'alternative sciences' seems to have put the clock back on the hard-earned progress modern science has made in knowing how to know. Moreover the 'local' knowledge systems they speak of in connection with 'Indian modes' of knowledge are not consistent with each other and more often than not 'patently irrational, obscuran-

[29] Noretta Koertge, 'Review', *British Journal of Philosophy of Science*, 50, 1999.

tist and downright oppressive to the same subaltern on whose behalf these intellectuals claim to speak'.[30]

In a heterogeneous and multicultural society like India we cannot afford to dispense with the notion of reason and science which is basically the spirit of the Enlightenment and has become a much maligned target of the fashionable post colonial post modernist critiques. But in an open society that admits of a plurality of interests, identities and self chosen ends (at least our constitution promises that) it is important the public sphere and the norms of discourse are not guided or dominated by any *one* class interest. Public reason must be conceived of as a never ending critical debate among contending interests which can be judged through publicly testable experience, independently of the class/ caste /gender identity of the arguer. This is the fundamental point of critical rationality in tune with the Enlightenment ideal which is indicted by various postmodernist critiques of modern science and epistemology. This ideal of rationality was not unknown to the architects of our constitution .It is reflected by the way the rights and privileges of the individuals are enshrined equally in the constitution irrespective of caste and creed, gender and religion. Equality presupposes rationality; it ensures equal rights for everyone irrespective of all social differences.

The culturalist understanding of science and knowledge has diverted people's science movements into indigenist and economic critiques of imperialist and colonial science rather than a self- critique of dominant and oppressive traditions. Once one takes refuge to a local or 'standpoint' epistemology all kinds of superstition /prejudice shall find a way to claim justification. The various forms of social constructionist theories of knowledge and science can slide into irrationalism of different sorts that we fought against for a long time. Human freedom as we know is circumscribed by social and political customs/norms/laws etc. so that other human beings' similar freedom is not violated. Thus the concept of human freedom is tied up with the notion of non-violence, for any amount of coercion amounts to some form of violence. 'Tolerance' like 'coercion' is associated with 'violence'.

The point is that the assumption of violence being a form of coercion only against human being is unwarranted. Why should violence basically mean violence to human beings directly (life) or indirectly (property)? The reason seems to exploit the standard appeal to the concept of 'respect to other human being', 'fundamental rights' 'obligation to posterity etc. I shall try to argue here that the standard notion of 'freedom' is too circumscribed and unjustifiably anthropocentric with special reference to various ecological discussions and theories. The present discussion will also aim to establish that the arguments which are used to defend human freedom can be extended not only to other living creatures but also to defend the basic freedom of the

[30] Meera Nanda, *Breaking the Spell of Dharma*, p. 119.

rest of nature. To do that we shall turn to the current debate on ecological issues and their various proposed solutions and conclude that all of these theories (and practical proposals) are based on reasons that take basic human concern as the parameter and fail to go beyond it to nature as such.

## *Tradition and Green Policy*

In a wide sense 'green' refers to general views and attitudes in the main part inspired by the international environmental/ecological movements, although the authors and activists differ in their styles. One thing at least is common: the disillusionment with the so-called scientific/industrial civilization, which ignored and suppressed various non-industrial traditions. From a rather narrow point of view green policies are characterized only in terms of pollution, resources and population control.

In a previous chapter 'Tyranny in Disguise' we have already discussed in details how scientific/expert knowledge and the related industrialization dominate modern life ignoring traditional modes of knowledge that reigned for thousands of years. Feyerabend has argued that the so-called 'rational debate' is not open since it is guided by scientific reasons, an open debate is one guided by pragmatic philosophy. 'An open exchange respects the partner whether he is an individual or an entire culture, while a rational exchange promises respect only within the framework of a rational debate'.[31] What he probably wanted to emphasize is that any encounter between industrial and non-industrial traditions or cultures is largely determined by the superior power of the former. In any kind of debate and decision-making, characteristics of the powerful prevail. Feyerabend cannot be more right in this premonition. In innumerable research and projects undertaken by various national and international agencies on ecological problems - the general picture emerges one of conflict and confrontation between traditional, weaker and poorer people on the one hand and scientific/industrial, stronger and prosperous bodies on the other.[32] Although Feyerabend's use of 'rational' to refer to scientific and 'non-rational' to 'traditional' is not undebatable for the traditional decision-making is not purely devoid of reason and in fact from their practical success seems to be quite rational. As we will see later from various accounts, traditional decision-making has been quite pragmatic and people in non-industrial period have survived for a long time without industrial technology. Even now non-industrial traditions - tribes and forest people can

[31] P. Feyerabend, *Science in a Free Society*, p. 131.

[32] In a self-ironical way Feyerabend commented on the occasion of buying his first car 'I have joined the eco-criminals'.

cope with their problems reasonably well without industrial aid.[33]

The two key notions of ecological theories or green philosophy as it is called are 'maximum diversity' and 'minimal interference'. And both these concepts are based on the notion of tolerance as suggested in the beginning. As originally envisaged by the libertarian principles plurality and diversity requires freedom. Maximum diversity of life forms, functions, environments, traditions, practices has to be recognized as basic to change and particularly to evolution of new forms of life styles and cultures. Moreover it is politically required in to-day's overcrowded and fast changing world. The so-called scientific/rational model seems to give a general sweeping standardized account of problems and practices. But cognitive diversity is not impossible - already the various non-scientific traditions have tried to solve various problems by variety of practices - some of which are successful and some are not. But then even scientific/industrial practices are not without some failures; and the cases where there are successes, many unforeseen adverse consequences were detected which tends to make the success itself questionable or at least raise the question whether it is worth. Anti-biotics for example have cured many difficult diseases, while their adverse side effects are notorious. Heavy industrialization changes economic condition quickly, but inflation, unemployment and sociological problems also follow it.

However the problems mentioned above are neither major nor peculiar only to a particular tradition. The more important point is that if diversity is required then can non-scientific tradition provide us with that rather than science or technology with one unique standard of rationality? Unfortunately the answer is not a straight 'yes'. The issue is seemingly paradoxical. Rational/liberal paradigm is clearly in favour of change since it is based on reason and reason only. On the other hand traditions are many and diverse, and as these are based only on practices, traditions seem to be insular and anti-change. We also witnessed history of intolerance (often violent) of one tradition (usually more powerful) with another (usually weaker) tradition. So diversity is not typically encouraged in traditional cultures which resent change at the beginning and only slowly accept it. The other basic principle the green philosophy is based on is minimal interference with nature. This is again a modest attitude of traditional cultures because of ignorance of effects. The assumption behind this principle is that natural phenomena are very complex containing so many variable factors that it is difficult to say what factors will bring what kind of change. Many ecologists prefer to keep interference with nature to the least for we are not quite sure about the future

[33] Shekhar Pathak, an expert on Himalayan ecology, has shown a picture where the traditional Himalayan tribes diverted a spring in such a way as to supply drinking water. The government has put a tap nearby which does not work anymore but the original spring still does.

effects. For example 'one cubic foot of good soil contains hundreds of species of organisms, each represented through hundreds, millions or even billions of specimens. They interact with non-organic ingredients in a more complex way than can be grasped in detail. But details, or specialized complex functions, are often decisive for the function of the whole as 'good soil'.[34]

What is applicable to a possible interference in nature is also equally applicable to non-industrial cultures. Many of the "developmental" projects are nothing but benevolent interference in these traditional non-industrial cultures. These would be paternalistic measures in their freedom, and big dams, factories or other such 'gifts' would not only be unwanted and unnecessary for their life forms but may also have disastrous effects on their culture. Feyerabend envisaged a free society where sub-cultures within industrial society are possible where unfortunately any kind of sadist community can exist. But Mill's one single principle of liberty can forbid that because of its rider, 'a person is free so long as he does not bring any harm to anybody'. So tolerance with others is a basic principle, but it has its limits.

Least interference and tolerance therefore can be extended to protection of animal lives and animal societies in addition to protection of human lives against undue interference. This is however complex and needs lot of normative analysis whether protection of one animal society (say the sheep) from another (say the tiger) would be paternalistic interference or not. More so when one party is not domesticated but 'natural' prey of carnivores. Because carnivores and antelopes for example are traditionally related as predator-prey order, so the 'minimal interference' principle requires that humans do not destabilize the natural order. The Bacon-Descartes ideal of cognitive superiority seems to justify interference in the name of understanding better. But respect and dignity for others encourage non-interference or least interference either in case of an individual or a community as a whole.[35]

## *Tolerance and Freedom*

It seems from the two principles stated above that both of them together advocate the basic norm of tolerance or 'live and let live' maxim. In this connection the libertarian account of freedom and reason may be taken into

---

[34] Arne Naess, 'Paul Feyerabend – A Green Hero', Munver, ed., *Beyond Reason.*

[35] In the chapter 'Tyranny in disguise?' it has already been mentioned how the non-Western traditions respect the dignity of human body in medical diagnosis by least interference while Western medical science undertakes all kinds of interference with the human body for diagnosis (and treatment) many of which are painful, expensive and not necessary.

consideration. For Mill and many others freedom and reason vary directly. Indeed it may be questioned as Feyerabend did, why what would be rational is to be decided by the so-called experts and not by popular consensus. Be that as it may, the moot point is: are the limits placed upon tolerance and freedom by Mill defensible or freedom and tolerance should be unrestricted to avoid any misuse? Feyerabend especially in his *Science in a Free Society* proposes the latter kind of unbridled tolerance. To him a society based on rationality is not entirely free; 'one has to play the game of the intellectuals'. In brief, he thinks the tradition of Western Rationalism/Science is inimical to a free society because (as has been already discussed before) the experts of a given tradition - which wrongly perceives its own tradition as situated outside all other traditions as an archimedean point of reference. As a result the tradition and its "experts" use chauvinistic and paternalistic tactics to suppress liberty of people and run society. This attitude also promotes a kind of Western imperialism over other cultures. It imposes a rationalist despotism and demands a conformity, which denies the basic tenet of freedom. For objective knowledge, he thinks, in quite Popperian way, a variety of opinions are necessary. And indeed 'a method that encourages variety is the only method that is compatible with a humanitarian outlook.'

*Tolerance and Paternalism*

Tolerance in principle is opposed to any form of paternalism. The latter is again an offshoot of a monolithic conception of truth and rationality. Feyerabend and a large number of eco-philosophers now distinguish between the participants in a tradition and outsiders. The important point to be noted is that members of a tradition have a right to run their lives as they see fit. A tradition with strange beliefs and even false beliefs should be tolerated. This is not to encourage right to have wrong beliefs or espouse anarchy but to recognize diversity and the principle of least interference. Is intolerance then admissible if the way of a society is violence and not tolerance for example? Mill's harm principle as we see will take care of the problem because the limits of tolerance there is the limit of 'no-harm' to others. But Feyerabend's free society would allow violence as a way of life so long as it is restricted to that society. It is not clear whether that kind of society is available now except in remote tribal habitats, for communication revolution already made different cultures closer to each other than before. Moreover people from different kinds of tradition often have to interact with each other, for example the government or non-government workers and the forest dwellers need a lot of interaction. It seems either there would be confrontation or persuasion which is bound to be somewhat paternalistic. The only move would be to follow the path of least interference, i.e., traditions or ways of life should not be disturbed as much as possible or unless the reason for change is overwhelmingly greater. But this is the Millean way and it falls back upon the concept of 'reason' and finally to some decision making process outside the tradition. This move is far from the Feyerabendian concept of free society and smacks of an irreducible element of paternalism.

## *Freedom and Happiness*

Most green philosophers would accept the Feyerabendian motto of freedom, tolerance and happiness rather than truth, rationality and knowledge as of intrinsic value. In other words knowledge and truth should serve humanitarian interests and not the other way. Human happiness is the primary principle and truth and knowledge are only secondary or instrumental to it. A free society is 'a society in which all traditions have equal rights and equal access to the centres of power' and traditions are also inescapable, so all traditions need to be supported to allow individuals within them more freedom.

This picture of tolerance of all traditions however seems to be acceptable only superficially. First of all although all ecophilosophical theories seem to accept that every traditions should be respected and therefore tolerated, the picture is both Utopian and simplistic, Utopian because the picture of peaceful co-existence of different traditions is unrealistic. Real life tradition seems to differ from one another and very well be clashing with each other. Moreover scientific/technological advances or at least changes are bound to have an impact not only on these societies but also on the individuals within a particular tradition. The picture is simplistic because we assume that all these traditions can be equally respected and tolerated. It overlooks that many traditions themselves inbreed despotism, tyranny over the weak and encourage violence. For example first, if one of the hypothetical tradition believes that 'kill all the members of the other traditions' or 'rob the outsiders of all their property' - do we still stick to the principle of tolerance and non-interference? Secondly, the concept of human happiness is very complex and sometimes theoretically loaded.

As already taken up by the utilitarians, it is difficult to measure the index of happiness. If traditions were tolerated as such then the criteria of happiness would be different in different societies except in some basic areas of human nature. Moreover 'desired' and 'desirable' should be distinguished in a wider moral context. If a person (circumscribed by her place in a tradition) is happy with her position that does not imply that it is morally desirable. For example if a twig-gatherer in a village near a forest spends her whole day in walking a long distance and have a frugal meal in the evening only after the whole family is fed; with no education and amusements but quite happy, should we say that it is a desirable life? She may be happy because there are no ways for her to know or desire for a different kind of life, a better life. So 'happiness' would be an empty concept (unless all of us aim at spiritual happiness or nirvana) if the concept of standard of living/quality of life were not recognized as an important parameter of happiness.[36]

But despite these internal difficulties we can nevertheless appreciate the spirit of a proposed free society. Feyerabend *does not* say that a free society will have *no* enforced standards. All he says is that the standards will not be epistemological and should be democratically arrived at, not imposed by one tradition (usually the more powerful one) on every other. The conception of a minimal state such as Nozick's or the notion of participatory democracy can bear out the basic tenets of Feyerabendian free society in moderate terms. But the kind of Protagorean relativism this view preaches and the role of intellectuals it perceives are not acceptable. If this kind of relativism 'is *reasonable* because it pays attention to the plurality of traditions and values' and " *civilied*" because it does not assume 'that one's own village and the strange customs it contains are the navels of the world' - then we need to change the notion of "reasonable" and "civilized". If plurality of tradition is to be recognized then it has to be done from a meta-level. And use of "civilized" (what is wrong with 'rational'?) again is complicated - for no traditions as such regard other traditions as equally plausible unless we talk of a super-tradition from above or at least outside the traditions which are under review.

But the recognition of such a super-tradition is exactly what Feyerabend and hordes of relativists would not allow. About the role of intellectuals it is simply given the status of 'just one tradition'. Their views are of no special interest or value. Neither are they capable of solving any problems nor is there any need of abstract ideas or philosophy to guide the solution of problems. I have already argued in details the problems of getting rid of the experts or intellectuals in a society but one thing at least should be mentioned here that usually by *intellectuals* we mean the people who can transcend their particular intellectual/cultural tradition and show the capacity to judge different traditions, theories, etc. from outside the traditions. People who cannot transcend the traditions they belong to (with their norms, bias and values) cannot be regarded as intellectuals, nor can they judge different traditions, norms etc. Traditions are judged by their own norms, but surely the norms themselves are not beyond scrutiny.

---

[36] Amartya Sen gives prime importance to 'standard of living'/quality of life for any economic ethical norm.

## *Human freedom and Nature*

We have seen that recognition of different traditions is only necessary to understand the concept of freedom. But diversity of forms itself does not guarantee that they cannot be adjudicated nor that freedom is restricted only to humanity. The awareness of animal rights and other ecological issues are gradually gaining ground to have a better understanding of the concept of freedom as we will see later. It seems that ecological arguments defending nature and natural habitat of people and animals usually exploit ideas like fundamental rights, respect to different cultures, obligation to posterity and so on. Most of the reasons are unfortunately anthropocentric-nature should be saved or different cultures tolerated for future human enjoyment or knowledge. Some species of creatures (or kind of culture) should be protected or preserved from commercial or other industrial type onslaught so that future human generation is physically, epistemically or spiritually benefited. Most ecological arguments use slight variations of this point. Tremendous amount of empirical findings, fieldwork etc. is supportive of a general anxiety about *our* future and *human* concerns. There is some occasional reference to animal rights but seldom we find an argument that considers nature as such to be respected and left alone as we argue for other fellow beings. To discuss these issues we now turn to the various ecological arguments and accounts that are put forward.

## *The Idea of Ecological Rationality*

Ecology generally stands for 'the man-nature ratio as a whole'. It is also felt generally that the symbiotic relation between man and nature is now at peril by mindless exploitation of nature in various ways. The main culprit is heavy industrialization and the commercial consequences thereof which are fast changing the natural flora and fauna of different terrain with their wild life and inhabitants.[37]

There are two strands of any philosophical analysis of the environment (i) exploitation of nature as such, (ii) effective planning of nature. The first one divides it into different specific components like biotic, abiotic, their relations etc. The second one consists of the politicization of the environmental debate, scientific and technological progress that is perceived as the reason for the destruction of environment. Both these aspects can be defended by the principles of maximum diversity and minimal interference with nature. The second is a milder form than the first but nevertheless not defendable in most

---

[37] Herbert Horz, 'Ecology as a Challenge to Philosophy', *Beyond Reason,* ed., Munver.

cases. Management is a better word than destruction but still involves interference in every possible way.

## *Rational Ecology or Philosophy of Nature*

An inevitable question for a philosophy of nature to address would be why should we respect earth's natural system and follow the principle of least interference and maximum diversity? This is a normative issue and should therefore refer to value of nature as a whole system.[38]

Even a non-specialist knows that nature gives us life-giving oxygen while plants absorb *poisonous* carbon- dioxide which other creatures and humans exhale. Earth's natural system *is* valuable for various reasons. It is first of all renewable, a regenerative process without any external or exogenous interference. Nature is also the genetic resource for human productive activity. It treats and recycles human waste products by gas exchanges as said in the beginning, responsible for supply of nutrient recycling water supply and purification and temperature control. The environment stabilizes ambient air and water quality. Many parts of the environment have pure amenity value in as much as they yield human enjoyment. Some people regard ecology as a source of values such as diversity, homeostasis and adaptiveness[39] or of a thoroughly non-hierarchical view of the world.[40] The latter is very significant for the 'hierarchies' in nature the ecologists speak of is entirely different from hierarchy in the sense of domination of one population by another, which cannot exist in ecosystem. There is hierarchy only in the sense of levels of organizations - from the genes to the eco-sphere, from unicellar organisms to highly complex systems such as human body, from ordinary grass to Himalayan perennials.[41] Self-regulation also is regarded as a value of nature,[42] the rationality of an eco-system is manifest in its quality of self-

---

[38] Ecological problems concern discrepancies between ideal and actual conditions stemming from interactions between human systems and natural systems. Ecological problems are characterized by (i) complexity, (ii) non-reducibility, (iii) variability, (iv) uncertainty, (v) spontaneity and collectiveness. See, J.S. Dryzeck, *Rational Ecology*, for such an account.

[39] J. Lemons, 'Cooperation and Stability as a Basis for Environmental Ethics', *Environmental Ethics*, 3, pp. 219-30.

[40] Murray Bookchin, *Towards an Ecological Society: The Ecology of Freedom; the Emergence and Dissolution.*

[41] See Ralston, 'Values in Nature', *Environmental Ethics*, 3, pp. 113-128 for a more extensive survey of reasons for valuing nature.

[42] Dryzeck, *op. cit.* The value of natural systems is critically analyzed also in. A. Leopold, *A Sand County Almanac*; M. Sagoff, 'On Preserving the Natural Environment', *Yale Law Journal* , 84, 1974. C. D. Stone, 'Should Trees Have Standing? To-

regulation, namely in its resistance to adverse conditions and resilience, i.e., its gradual adaptiveness to survive or overcome in changed condition. This in brief are some reasons to value an ecosystem as such.

### *Nature and Man*

As mentioned before in pre-industrial time man was closer to nature in the sense of accepting it as such interacting in a more direct way with no aggressive attitude towards it. In other words, there was a give and take relation with no more demands on it than the basic requirements. Nature in its diverse forms- mountains, river, sea, plants and other animals were regarded with respect (sometimes fear) and affection, this attitude is regarded by intellectuals as animistic, i.e. attributing spirit or mind to inanimate objects. It was common practice not to milk a cow for quite a while after the calf is born, or not to tear leaves or fruits from trees when they are supposed to 'sleep'. All these attitudes are changed by the so-called epistemological conquering of nature by science, which claimed to 'know' and dominate nature so that the latter can serve human beings by its enormous resource potentials. Thus the symbiotic relation in which nature and man lived for millenniums have been snapped by the onslaught of western science only for about less than three hundred years. And in this short period nature has been destroyed, depleted, exploited in such a degree that our five thousand year or more old civilization could not think of. The worst part of such conquest, most people feel is that the majority is not benefited by it, only a minority privileged class became more prosperous and beneficiary to the various modern facilities while the people who lived for thousands of year drawing sustenance from nature if not happily but at least in peace were made suddenly to accept the fact (by government regulations) that they do not have any right to use nature- forest, river some land because these are not their property.[43] The story of such exploitation of nature is similar all over the world but it is more poignant in the colonized countries which were rich in natural resources because the resources were ruthlessly exploited to make the colonial powers enjoy the

---

wards Legal Rights for Natural Objects, *Southern California Law Review*, 45, 1972; L. H. Tribe, 'Ways not to think about Plastic Trees: New Foundations for Environmental Law', *Yale Law Journal*, 83, 1974.

[43] In India the British foresters were specially used for implementing such imperialist policies and this started the confrontation between forest officials and peasants, mainly marginal farmers. Forest acts were enacted to deprive these forest-dependent people from the use of the forest that they survived on for years. See, R. C. Guha, ed., *Social Ecology*, and R. C. Guha, *The Unquiet Woods: Ecological Change and Peasant Resistance in the Himalayas*.

fruits while the indigenous people were pushed to more poverty and indignity.

Apart from the specific account of exploitation of nature by the colonial powers, the exploitation in the name of industrial development, many ecologists think, are not only wrong but also utterly myopic about its disastrous future consequences. In other words any sensible mind will question the rationality of 'developmental' work that overlooks the adverse effects of such project. So we need some principles of ecological rationality. If the principle has to ensure rationality then even if it requires a degree of intervention (interference) in natural systems, it should discourage extreme ecological engineering. Man can make use of rather than seek to supplant the spontaneous self-organizing and self-regulating qualities of natural system.[44] An ecologically rational man-nature system is one in which human and natural components stand in a symbiotic relationship. We cannot restore everything that is lost but at least the awareness, the active and sincere attempts to make up and pledge to minimum interference in future and protect maximum eco-diversity would be rational. This as a principle is abstract, but various eco-movements to save nature have blueprints for actual implementation of such a principle of rational ecology. The implementation is necessary because after all man cannot exist without a supportive nature but nature can exist without man despite some improvements undertaken by human beings for self-interest. In the long run only species that affect their environment positively can survive.

Besides this, the rationale of the basic norm of 'live and let live' corresponds in terms of traditional philosophy, to a norm of 'maximum fulfillment of life potentials'. This norm implies norms of diversity, complexity (of structure and function) and of maximum symbiosis, (as mentioned before) that is arrangement of life forms and life styles with minimum negative interference with each other. An interaction counts as negative (as hinted in the final sentence of the last paragraph) if it decreases the potentialities of life fulfillment of the participants.

## *Social movements and Nature*

### *Nature and social practices: an ecological perspective*

As mentioned before although there is hierarchy in nature, that hierarchy is not domination of one species by another, symbiotic relation means, diversity complements various species to an overall supportive system. But different social practices are going on after the nomadic days of hunting and gathering

[44] Dryzek, *op. cit.*

is over slowly but steadily altering the environmental conditions. Tilling, forest clearing, indiscriminate stock grazing and later intensive farming took place in such an extent that a large number of plants and animals failed to survive the onslaught. Again by various means man has improved, augmented both yield and variety of crops, plants and his daily nutrients. In brief the frontiers of natural produce and cultivation has been extended. But these augmentation and extension processes are *not* required by nature, nor the latter is 'enriched' by the factory-produced chemicals or tools. The different social practices classify and evaluate different natural things as precious or not. Nature does not have any standard to do that. For example the so-called natural classification is also socially determined, why for example water hyacinth or weeds has to be destroyed while corns and sugar cane protected? The reason is obvious that whatever is suitable for human consumption and at least enjoyment (e.g. rose) is to be preserved while the rest should be destroyed or at least minimized to make room for human need-fulfillment. The broken balance between man and nature thus creates an uneven and paradoxical picture. Certain species of creatures[45] started breeding in such condition than in wild nature. For example in the crop fields insects, pests, parasites grow through the unlimited supply of food which is often difficult to exterminate.

Most ecologists due to their humanitarian outlook seem to forget that human beings are part of nature as much as animals and plant communities are. Ecological theories are generally concerned with biotic factors with the effects of any natural things upon man only. Indeed sometimes there are oblique reference to animals and plants but that is also with *human concern* in the end. An 'undue importance has been given in history and economics to these purely human influences.' Like other social deterministic theory most social ecological discourse seems to ignore that if we talk about symbiotic relation with nature then each community there should follow similar rules that maintain a balance and rhythm of growth for all. Mukherjee maintains in similar vein that 'each community cannot appropriate more than its due place in the general ordering of life from which nothing can be obtained without influencing everything else. Working symbiotically, they represent interwoven threads of a complex web of life'. No threads can be snapped or removed "without the whole garments of the life and nature and human society being disfigured".[46] The old fashioned metaphors used by him to describe a hypothetical state of affairs less than seventy years ago seem to be a realistic description today.

---

[45] Raddhakamal Mukherjee, a pioneer in ecological research in India as mentioned before gives examples of disbalance man brought in nature. He says, 'The plants which he (man) does not tolerate are called weeds', ibid.

[46] Radha Kamal Mukherjee, ibid., p. 26.

*Nature and human society, some experience of conflicts*:

*Man Nature conflicts*

Although this kind of conflicts can assume many forms and history of human civilization is one of constant conflict, compromise, defeat and conquer - we shall restrict only to some empirical accounts to understand some basic point about this relationship and how it has changed over the years. I have already mentioned in a footnote before, that in India with the advent of British rule the prudent indigenous practices of maintaining stability in natural resources and human need-fulfillment, the symbiotic relationship was upset and so was the Indian society. The British imposed highly demanding claims on natural resources of the country to supply raw materials which are rich in quality but cheap in price for their economy both in India and elsewhere in the Empire.[47]

The standard method of this practice was to declare vast resources (land, water resource etc.) which were used and owned communally till then - as government property. Instead of poor villagers and forest dwellers using the resources these were commercially exploited in a large scale. This tradition not only continued after the British left, but the trend of rapidly depleting natural resources commercially has accelerated over the last three decades since independence. This has led to considerable impoverishment, and often complete collapse of the natural resource base sustaining many components of the Indian society.'"[48] The prudence of the earlier traditions were evident in many practices such as regarding some place, tree or animals as 'sacred' or forbidding plucking, tilling, or harvesting in some period of the year which helped preservation and stopped excessive exploitation of nature. Modernization and government take- over destroyed the traditional practices with no rational alternative in conservation principle.[49] Adaptiveness or resilience as a principle discussed earlier thus is being lost by the commercialization of natural resources in India.

Various field reports and research have supported the view that it is a misconception to hold that marginal farmers and forest dwellers are respon-

---

[47] The valuable anthology *Social Ecology* Oxford 1994 edited by Ram Chandra Guha contains various informative accounts of such ecological problems in India raised by the modernization processes started by the British and followed thereafter. The modernization process (with its numerous 'forest laws') gradually depleted natural resources to a great extent and marginalized traditional forest dwellers and forest users.

[48] M. Gadgil and K. C. Malhotra, 'The Ecological Significance of Caste', in Ram Chandra Guha, ibid., p. 37. The authors have argued that the collapse of the resource base, the change of barter to money system has destroyed the traditional relationship amongst the various castes in Indian society.

[49] Some examples of such traditional prudence and their loss by modernization are to be found in Gadgil and Malhotra, ibid.

sible for depletion of natural resources. Even those who depend on forest and other natural resources (like river, lake, etc.) take very little compared to the huge amount of commercial exploitation. Moreover their need is very small and inexpensive while the commercial depletion is rapid, drastic and difficult to be regenerated. An analogical misconception is that the poorer countries are responsible for global pollution. The fact is that the so-called highly industrialized countries contribute to more pollution in the form of physical and chemical waste of their industries, automobiles etc. than the less industrialized ones. For the gradual tightening of the pollution control laws however in these countries, a new strategy is evolved to shift these polluting industries to third world countries in the name of globalization. For the somewhat non-rigid law and implementation practices there, aided with prospect of new job opportunities - the wastes of these countries are shifted to the third world while the profit and prosperity remain there.

To come back to our original point of man-nature relation, due to the changing government policies the native dwellers and users were pushed to utter poverty while the commercial exploitation made a minority class very rich. As mentioned before the native users of natural resources were never very harsh with nature since they live with it in a symbiotic relation for generations while the commercial users are outsiders and mercenaries, so their treatment is purely professional and drastic. Management and control of forests created a chasm between man and nature. The so-called 'reservation' of huge area of forest created an alienation of humans from nature and the latter became associated with the government and thus perceived as an adversary.[50] The peasant movements against commercial forestry, large dams, and mining are consequences of such alienation from nature. The 'reservation' meant an effective loss of control over their habitat for forest-based communities. A fundamental change has taken place in the agrarian relations induced by (social) scientific forestry. Through the mechanisms of the forest acts, the state preferred to deal directly with individual households (rationing their requirement) rather than with village communities which is a greater threat to commercial forestry. The loss of community ownership leads to have broken the link between humans and the forest. The practice of having limited tracts of forests, the so-called 'village' or 'third class forest' for public is to encourage preservation of tree by people. All these different policies and practices gradually lead to erosion of social bonds and Guha thinks, as a result the alienation of humans from nature would be complete.

Similar 'melancholy effect' forest reservation had on the tribals of Central India[51] for taking away the forest they regarded as 'their own property'.

---

[50] See, Ram Chandra Guha, *The Unquiet Woods*, for an account of such alienation between nature and man in the hills of Kumaon and Tehri Garhwal.

[51] Verrier Elwin, *A Philosophy for NEFA*.

In Europe too the takeover of woodland for hunting or for timber production was deeply resented by the peasantry. Sometimes extensive forest-fire by peasants shows their hatred for the forests and their oppressors.[52] In its extreme form alienation occasionally forced peasants to degrade the surroundings with which they once lived in symbiosis. It is an irony that villagers in part of Garhwal look upon the reserved forest as their main enemy with the wild animals that destroy their crops.[53] Alienation thus leads to dehumanization and distorts the basic human qualities. The alienation and the growing animosity of the peasantries were also not unknown to our colonial officials. To contain the resentment the natives felt to be deprived of their freedom, the British foresters allowed peasant access to restricted areas not deemed commercially profitable while completely forbidding them to enter areas (with strong punitive measures for intruders) under commercial working.

As I mentioned before, there is no hierarchy in nature in the sense of more or less 'powerful', more or less 'precious' and so on. But a classification of natural products is being done on the basis of their commercial value, with no reference to their ecological function to the environment or relevance of the species to the forest people's need. Even human users were redefined as possible source of injury or as 'enemy of the forests' (who are the friends? the forest merchants?) in the same category as natural hazards and wild animals. Recent researches by Indian ecologists find a chasm between the ideology of scientific forestry and the actual operations of timber harvesting and a conflict is also noticeable between the imperatives of scientific forestry and the economic and cultural values of the hill peasantry.[54]

## *Man - Wildlife Relation: Conflict and Compromise*

The most difficult aspect of ecological balance and recognition of natural right are the questions of such right to wild animals. Forest and its human dwellers (even if some of them are ferocious tribes) do not pose such difficulties as wild life does. We know indiscriminate poaching and unnecessary hunting games and commercial hunting (e.g. elephants for their tasks, whales for their oil, mink, angora etc. for their expensive furs) led to near extinction of many species. Moreover as natural inhabitants of forest (unless they intrude and destroy people and their property including domestic animals) - why should not they have a prima facie right to live freely? Apart from this

---

[52] E. Weber, *Peasants into Frenchmen.*

[53] P. H. Gross, *Birth, Death and Migration in the Himalayas*. I owe the three references above to Ram Chandra Guha's *The Unquiet Woods.*

[54] Gadgil and Prasad, 'Forest Management and Forest Policy in India', *Social Action*, vol. 27, no. 2, 1983.

moral question, the empirical issue is also important, namely how best to adhere to the principle of conservation of wild life to maintain the ecobalance of nature?

In case of India traditionally the moral and the empirical questions are solved in an integrated approach of conflict and compromise. In other words unnecessary killing or confrontation was kept to the minimum, (adding to the symbiosis with nature) the only exception being the assault or offensive attack to their life or property. That means the two principles of diversity and non-interference are followed in case of wild life as well as inanimate nature. It is well known now that receding forests and depletion of natural smaller herbivores make carnivores and big herbivores (who feed on trees & plants) like elephants invade villages only because they do not get sufficient food in the forests. As a result wild life and peasant/villagers' conflict is an inevitable phenomenon, but the exploitation of the forest is indirectly responsible for such confrontations. In an alarming way killing of honey and wood gatherers in Sunderbans in Bengal by tigers is increasing every year. The animosity of the poor inhabitants of this region thus is increasing but still they need the forest for their living.

Despite these stories of conflicts and confrontations between humans and wild life, people in the Indian subcontinent have dealt with wild animals such as 'rogue' elephants or man-eater tigers through the ages in a reasonably effective and tolerant way. Otherwise these animals would not simply have survived to this day given the long history of settlement in the subcontinent and the large human population. Elsewhere people have often solved their problem by simply wiping out an offending species. During the nineteenth century for example, wolf, a predator on sheep in Great Britain was completely exterminated. In the past, religious taboos no doubt played a role in this tolerance towards other creatures. However in the present context of a rapidly changing society such traditional attitudes to nature are unlikely to sustain the conservation ethos for long. For it is unjust to expect only a certain section of society, the marginal farmers and tribals to bear the entire burden of conserving wild animals. 'Project Tiger' would be meaningless to a poor honey-gatherer in Sunderban whose family is killed or near- killed by tigers unless some viable scheme is proposed to gain acceptance among people who interact directly or indirectly with wild life in their daily lives.

From the perspective of this man-wildlife relation of confrontation the question we raised in the beginning of this section is inevitable: why conserve wild life? The primary argument as already mentioned is: (1) ethical: living creatures have an intrinsic right to exist irrespective of their utility or otherwise to human beings; (2) ecological:[55] large mammals such as the elephants or tigers play a dominant role in an ecosystem by virtue of their large bio-mass or position at the summit of an intricate food web. Disappearance or depletion of such key species could lead to the disruption of ecosystem functioning, perhaps resulting in extinction of other species; (3) economic: the direct value of wild animal's returns from wildlife oriented tourism and so on;(4) aesthetic/entertaining/educative: wildlife is a source of considerable pleasure for many people thereby having a therapeutic value in promoting human well-being.

These arguments are quite convincing to justify conservation of wild life although as mentioned before, unless a viable strategy is evolved, the marginal farmers and forest-dwellers who have to deal with them in their everyday life would not so readily accept the rationale.[56] However reasonable these grounds are, except the first one, all these arguments are instrumental. I call them instrumental because these are put forward only as a means to achieve some goal for human beings. Wild life is used instrumentally for fulfilling some human need - not for wildlife as such. Only the first one is treating wild life as an end in itself by asserting its natural right to live as any other living being. But the problem is: then bacteria or poisonous plants also should have that right at par with man and all other living things. One single rider can be proposed for such problem: every living thing has the right to live (in their natural habitat) so long as it does not override some other's similar right. It is clear that except this ethical argument for conservation all other arguments are anthropocentric, i.e. giving human being more value than others. But it is not clear why in an ecosystem human being would be in a more privileged position than others would.

## *The Concept of Ecosociety: A positive approach*

Although all ecological discussions go around the original man-nature equilibrium, the general attitude is one of despondency and frustration. The reason behind this negative approach (with few positive suggestions) lies in the

---

[55] For such arguments for conservation of wildlife, see, 'Wildlife - Human Conflict in India: An Ecological and Social Perspective' by R. Sukumar in *Social Ecology*, *op. cit.*, p. 304.

[56] R. Sukumar is aware of this and suggests that unless there is an acceptable policy, conservation would not be successful, the latter requires an assurance to minimize its impact on human lives and properties, Conservation can only succeed if the legitimate aspirations of people dependent on forests for their livelihood can be met by the rest of the society. If somehow the yield from wild life reserve/park reaches the local economy for example poaching would drop and village attitude towards wildlife/forest conservation becomes more positive. Ibid., pp. 314-15.

fact that big power and big money is involved in the industrial and so - called development projects that are the main actors in this destruction drama. Although the peasants are now united to some extent to rise against such power traditionally their use of forest was minimal and attitude towards forest was so long one of intimate relation.

The concept of ecosociety based on the principles of tolerance and non-violence is envisaged by many. These kinds of societies are usually more or less self-reliant, utopian[57] to a great extent. In India such communities are not new, 'ashramas' are not necessarily only religious places. Gandhi's experiments are living examples of a way of life as whole, united with the rest of nature in a harmonious way which modern industrialization has robbed us of. This is not a place to discuss how far Gandhi was actually successful in such experiments or the pros and con of such ecosociety that would deny the impact of heavy industrialization on people. But the basic tenet of such experiment is laudable, namely an ideal society is that where the essential bond between man and nature is respected and not interfered with as much as possible. This kind of hypothetical society is based on the two fundamental principles of maximum diversity and minimal interference. One can also argue for such rationale in the following way. Although man has subjugated nature in various ways and invented various implements, which made our lives easier in a way - people still are happier to see a natural wonder than a man-made tool. Or being a part of fast, industrialized world, human being still search for peace and respite in nature at least for a brief period to come back rejuvenated to face again a demanding modern life.

There are communities or eco-societies, which have actually thrived amidst nature and have been quite happy.[58] Hart's account (see footnote 28) is based on the arguments that happiness is a harmonious equilibrium between man and nature. Deeply influenced by Gandhi and also by the principle of non-violence of Buddhism the author founded a co-operative community on 40 acres of farmland and woodland in Shropshire Hills, England in 1982. He believes in self-sufficiency by Agro forestry - a system for supplying all basic human material needs from trees and other perennials grown in symbiotic association with each other. Citing examples of many societies that thrive in similar way, he thinks this kind of eco-society can be viable alternative to consumerist societies of modern industrial civilization.

This kind of society undoubtedly is a moderate type of solution to indis-

---

[57] The Utopian societies are isolated self-sufficient communities which usually do not interact with industrialized civilization as much as possible. The Mormons, Amish, Shakers and Pennsylvanian Dutch are such communities in North America. Typically such communities are bound by some religious faith.

[58] Accounts of some such societies can be found in, Robert A. De J. Hart, *Eco-Society: A Historical Study of Ecological Man.*

criminate misuse of nature to cater the huge demand of industrial development. But science and technology at present have reached such a state that it is impossible and unrealistic to completely deny its impact upon our modern life and culture totally. Isolated cases of such societies with consenting people cannot solve the basic problem of tyranny of 'development' over nature and people close to it. The best way is to evolve ways to keep the onslaught to the minimum and respect the rights of nature in a wide sense in a way as we wish our rights to be.

Most ecologists realize this and arguments can be given to show that it is more rational to regard nature with tolerance and least interference than not. The quantity and quality of diversity of biomass within any village ecosystem is necessary to meet the diverse household needs of fuel, fodder, building materials etc. and artisan. These are particularly relevant for Indian context and other poor countries, where most people still depend for their living on the whims of nature. In periods of emergency, such as flood or drought (and a resulting crop failure) which are regular phenomena - the various roots, leaves, even some wild animals became alternative sources of nutrition. In countries like India where most people still live in traditional ways, the combination of trees grasses crops, animals and ponds which we found in almost every village was a unique interactive and resilient system to natural emergencies.[59] The example discussed in the footnote below shows it is not enough to preserve biological diversity just in those areas where the flora and fauna are genetically rich and diverse by setting up biosphere reserves and national parks: we should ensure that biological diversity is preserved and/or recreated in every village eco-system. Social forestry programme or any such project would not succeed unless villagers are assured of security about their farming lands and livestock. Easy availability of biomass can bring desirable social change. An example[60] can be given of Kerala which is regarded as economically backward, but easy availability of bio-

---

[59] Not only in emergencies, this unique interactive system is also valuable for its regular support to urban demands. The neighbouring villages in the eastern fringe of Calcutta were an extra-ordinary example of recycle and supply, while a river there carries the sewage of the city. The dumping ground of the city nearby grows vegetable, in lands rich with organic compost. Several large fisheries (water bodies) there were well-known for their pisciculture. But for several years the fishponds are being filled up, the dumping ground minimized to make room for big housing development projects government offices and industries. In the past some tanneries were the only industries there. This change has two drastic repercussions: one, the villagers nearby easily earned their living in that naturally rich area by supplying provisions to the nearby city and the neighbourhood. Now their land/pond is overtaken somehow and thus they are turned into labourers and domestic servants. Two, the city and the neighbourhood now are deprived of cheap and fresh provision as they used to get before.

[60] Anil Agarwal, 'An Indian Environmentalists Credo', in *Social Ecology, op. cit.*, p. 375.

mass there free women from many hard and time-consuming chores. The highest literacy rate there shows women also find time to go to literacy classes while villages in Rajasthan, Gujarat are backward in literacy, especially women literacy as they spend much more strenuous time in fetching water, firewood, etc.

## *Social and Natural*

One way of understanding the interrelationship between society and nature is as we have seen to follow the principle of 'live and let live'. This is anti-paternalistic and based on the concept of basic freedom to live in one's natural habitat. The other way is to allow some kind of social engineering so that the ecobalance be ideal and the underprivileged is protected. The latter is paternalistic and it involves active intervention so that ecological crisis (primarily the human problems) are solved or avoided. The latter is advocated by Marx in his early writings and since followed by the 'green' social theory which perceives ecological movement as a protest movement of the peasants/forest dwellers against the profiteers of capitalism. For 'green' theory there is no purely ecological approach, explanations are couched almost wholly in terms of the activities of institutions such as multinational companies with their large wings around the world or with national and transnational governments. All these institutions, governmental or private, are becoming increasingly distanced from people's everyday lives and demands.

Environmental issues tend to be seen, within this paradigm as an integral part of a fundamentally exploitative, unjust and undemocratic form of society. "Natural resources are systematically depleted in the accumulation drive by both private multinational capital and the state. The costs of development are expressed not only in terms of class conflict and economic exploitation, but also in the reduction of the natural resource base on which the poor depend for their livelihood."[61] The multinational business, in this view is the prime culprits in wrecking society's resources as a result of indiscriminate forms of exploitation and production. This perspective certainly contains many powerful insights but it is linked to distinctive political programme. Moreover in countries like India, governmental institutions are equally if not more responsible for devastating nature with a shortsighted policy of making the two ends meet. Moreover socialization of the means of production as the programme suggests also cannot help much as we just said. It is also felt from within the movement that for environmental issues new forms of political programming is necessary although it is to be noted that the cause of

[61] M. Redclift, *Development and the Environmental Crisis, Red or Green Alternatives* ? p. 38.

environmental crisis is firmly anchored in the global industrial order.[62] Social change of a drastic nature is difficult to bring about while the ecological crisis calls for some immediate remedy if not complete solution.

The relationship between people's behaviour and their physical environment can be perceived more openly as done by the supporters of economic and political individualism (which concentrates more on the individual decision-maker rather than taking 'people' as a whole). It gives more importance to the individual and the household. As part of this it[63] draws attention to genetically based forms of behaviour and to the psychological dimension of environmentalism. Some neo-liberals also draw attention to the uncertainty of knowledge surrounding environmental issues.[64] These are helpful suggestions but encourage possibility of complacency. How can individuals plus the market force guarantee that once peoples economic conditions are better off and the market force ensure that natural degradation be stopped there shall not be environmental crisis, at least not for the people who have to depend on their surrounding for their living and sustenance? But the point is if there is really no need for public intervention, how can the exploitation of the 'common' resources be stopped or at least minimized? Moreover the 'man-environment' relationship is a very complex problem, while organisms are creative and make their environments in such a way as to become virtually

---

[62] R. Bahro holds that the old 'green' programme is not adequate to deal with ecological problems. See his *Socialism and Survival.*

[63] The liberal economist Hayek argues that man is both biological and of human nature. Contemporary human beings are still well adapted to the life of small hunting and gathering groups. But they do have special capacities like learning from their actions, avoiding purely instinctive actions. They have also evolved some rules of human conduct, based on values like honesty, morality, privacy, sense of property rights etc. which have allowed the human species to have acquired their special prominence in nature. Cultural relations of this kind, Hayek thinks now act as a constraint on human beings' genetic inheritance. The best kind of 'planning' would be carried out by intelligent individuals. The social and economic scenario is now so complex that no large-scale institutional planning (for the benefit of the people) can be done for lack of knowledge of the consequences. See especially, *The Fatal Conceit.*

[64] Some neo-liberals basically following the Hayekian line are skeptical about the environmental doom prophesied mainly by the green philosophers because the 'scientific' evidences for such dismal future such as ozone layer depletion, greenhouse effects etc. are still scanty and inconclusive. It is also doubted whether there are indeed strong links between changes to the ozone layer and the use of chlorofluorocarbon. Variations may well, for example, be linked to variations in the solar cycle. Likewise the links between skin cancer and ultraviolet rays or the belief about rain forests and the disappearance of the species are not established. Nevertheless the fact that the problem of pollution is real has to be admitted by them while as a solution they refer to capitalist methods, the green theory blames capitalism for the environmental destruction. For such view, see, R. Whelan, *Mounting Greenery*, 1989 and *The Limits of International Cooperation*, 1990.

part of it themselves, at the same time environments (nature and other people) are active in the making of organisms. In many respects organisms and environment-both the elements form part of one another. It seems we need some new framework, a unified approach to overcome this dichotomy. That needs[65] a few sets of epistemological concepts and value perception.

There are problems in either accepting that the exploitation of nature qua-people dependent on it is due to only capitalist economy (the problem is much more complex than a single cause) or conceding to the view that no social intervention is necessary, market economy will cure the malady and that the problem of exploitation of nature and the imminent doom is overstated. The first one is simplistic and the latter utopian. Nobody can deny the impact of industrialization in pollution and purely economic solution is not the total answer to the environment problems. It requires more than that. One very important point however is to note that in Hayek's account we find an argument for regarding man as something special and different from the rest of the nature. In the so-called green theory there is no special reason given why environmental problem is basically a social problem involving human being. Hayek's view that man has a special position in nature may be contestable, but he realized the need of a theory why we should consider man having a unique position in environmental issues. The view is contestable because one may argue that man's role in nature is not unquestionable for the balance of it is more tilted towards destruction than preservation.

As I mentioned before science and technology has done many mistakes, Popper echoes this view by admitting the failings of science and technology, nevertheless declares at the same time, "But to attack science and technology as a whole, when they alone permit the necessary corrections to be made, is not only stupid but sensation-seeking and completely irresponsible."[66] Popper's worry against too much legislation against technology is based on his concept of free enterprise which would be threatened by this policy. Although he admits that many problems such as air pollution may require special legislation. But he is opposing 'ideological worshippers of the so-called free market' who regards any kind of legislation limiting market freedom as a dangerous step towards serfdom.[67] This attitude again is an ideological nonsense to him. It shows as I mentioned before that Mill's single requirement of "harm principle" which requires freedom so long as it is not detrimental to others similar freedom and does not cause any harm –can handle the problem of freedom vis a vis development. Popper was very vocal of the imminent

[65] Peter Dickens both criticizes established theoretical approaches and argues for a new 'green' social theory based on the early writing of Marx as an integrated approach. *Society and Nature*, Chapter III onwards, contains such a view.
[66] Karl Popper, 'Thoughts on History and Politics', *All life is Problem Solving,* p 101.
[67] Ibid., p. 101.

danger of loss of freedom in the ideology of Green movement so much so that he declares that "Better to be Free than Green'[68] and does not seem to realise that the single rider of harm principle can take care of individual freedom as well as technological developments. This principle, namely not to cause harm or interfere with nature so long there is no overriding reasons for such violation. This principle is enshrined in the two principles of rational ecology. These are the principles of maximum diversity and minimum interference which can very well preserve individual freedom in an extended sense and at the same time is not detrimental to all technological development and progress. In other words one does not have to oppose *all* scientific and technological projects to thwart natural devastation and atmospheric pollution at least in principle. Popper thinks that every organism is constantly preoccupied with attempts to bring some improvement over their environment. But that does not mean science has destroyed that natural process. Indeed there are some unintended adverse consequences of scientific activities, but that cannot justify opposition to technological innovations.[69]Man's social nature prevails over natural and that brings us to the final section of this discussion namely the question of right of nature with or without man.

## *Freedom of Nature and Rights of Man*

The relationship of man with nature is such that as we observed before, nature can survive without man but man cannot survive without a supportive nature or 'biosphere' as it is called now. Therefore by protecting nature, man is still first and foremost protecting himself. The environment is endowed with no intrinsic value here, rather the action stems from an awareness that by destroying his surrounding man may be endangering his own existence or at the very least, depriving himself of the conditions for a good life on this earth. Thus nature is taken only indirectly based on a position that may be called 'humanist' or rather "anthropocentrist". Here nature is thus only human environment - the periphery and not the centre. As such therefore nature cannot have any legal right as it does not possess any intrinsic value.

Another way is to attribute moral significance to certain non-human beings. This move can be supported from a somewhat utilitarian principle according to which one must not only look out for man's best interests - but more generally try to both diminish the total suffering in the world as much as possible and increase the quantity of well being. This point of view goes

[68] Karl Popper, 'Meglio liberi che verdi', in an interview by Sebastiano Maffettone, GRANDI FILOSOFI/POPPER GIUDCA LE UTOPIE CONTEMPORANEE, *Panorama*,16, LUIGLIO 1989.

[69] Karl Popper, *In Search Of A Better World, vii-ix.*

beyond the concept of human freedom and is the basis of animal liberation. All beings capable of feeling pain and pleasure must be considered legal subjects and treated as such. It avoids anthropocentrism to some extent by including animals within the sphere of moral considerations.[70] Against the humanist anthropocentrism in ecology a more radical alternative extends this notion of moral and legal right of nature in itself including its vegetable and mineral forms. It is becoming a dominant movement in Germany and the United States and questions the issue of humanism in radical terms.

In this approach the old 'social contract' devised by philosophers should give way to a 'natural contract' in which the entire universe becomes a subject of law. As I proposed in the beginning, most ecologists (leftist, rightist or centrist) argue from the standpoint of human self-interests as ubiquitous from even the terminology they used. But only if the problems are perceived from a deeper and transcendent standpoint then we have a better understanding of the relation of man *in* nature and not man *and* nature, for man is just one part of nature and we should not overstate his position.[71] The concept of freedom that goes beyond man may be termed 'ecocentric' or 'biocentric' in contrast to the standard anthropocentric view popularly associated with green theory or other social theory of environment. A coherent doctrine of nature as a new legal and moral subject can be constructed from this perspective. But all the three levels of the issue have at least one thing in common namely their anti-modernity, i.e., arguments against the impact of technology and consumerist attitude created by Western scientism. One possible aspect or difficulty of regarding man only as natural is that although man is biologically and socially determined by a body and a physical situation and lives within a social milieu, a culture and language - he is nevertheless a free being, not an object. Because he can become very different kind of person by that freedom while the rest of the nature does not have equal capability of utilizing their right to develop into variety of beings. It is a very deep question and it is not possible to discuss here the different gamuts of this question fully (that requires a full-length discussion). We shall instead turn to the legitimacy of the anti-anthropomorphic view of nature and its rights.

Historically the perfect model of antropocentrism is found in Cartesianism which grants *all* rights to men and none to nature including *animals.* Not

---

[70] Jeremy Bentham is one of the pioneers in recognizing animal right and at present Peter Singer advocates it. Although Singer calls it 'animal liberation' rather than the question of right they agree on the essential point. See his *Animal Liberation* . 2nd edition.

[71] Besides Singer there are Aldo Leopold (U.S.A.), Michele Serres (France) and Hans Jonas (Germany) who are defending natural rights against anthropocentrism in different ways. Leopold's *A Sand County Almanac*, 1949, is a pioneer work while Serres' *Le Contract Naturel,* Paris, 1990, and Jonas', 1979, support a crusade against humanist ecological arguments and argue to go beyond. *The Imperative of Responsibility.*

only is nature not 'animated' but it also contains no occult power or invisible reality. And animals are no exception because they do not speak. Indeed they sometimes function better than men do, but so does well-oiled machines! They do not have any *rights* because they do not feel pain. Descartes nevertheless made a distinction between animals as creatures of God and machines fabricated by ordinary men. Animals still are automatons; internal organs intended for different functions just as a machine (say a watch) has internal parts for functions. The central argument[72] for animal right is clear; the specific differences ordinarily invoked to place value on human instead of animals (reason, language, culture etc.) are irrelevant. Clearly no greater rights are granted an intelligent/cultured man over an idiot. The only significant moral criteria should be the capacity to experience pleasure and pain (which Descartes denied of animals, to him even the cry of an animal from an injury is like the ticking of a clock).

The argument for animal right is basically one prescription in a democratic framework. The idea that no greater rights are granted to a wise man than to an idiot is the heart of a democratic worldview. So with or without a utilitarian thought namely the goal of moral and political activity is the maximization of the *happiness* of the world (of not man alone) and not primarily *freedom*, therefore to protect interests of whoever the subjects are. It is therefore just as wrong to bring suffering to an animal as to a human being. Peter Singer seems to go for the latter, but I think just the Tocquevellian requirement of increasing equalization of conditions in democracy is sufficient to argue for protection of animal rights. Singer considers (along with many others) the animal as worthy of respect in and of itself. He takes animal to be a 'moral subject' endowed with intrinsic dignity. He makes[73] a difference however between the sentient and insentient to include only the former under the privilege of having rights excluding the rest of the nature like trees, vegetables, mountains and rivers from the boundary of rights. So this kind of account[74] is only half way to understand the obligation of all sentient crea-

---

[72] Originally proposed by Bentham this kind of arguments is however used in more or less the same way by others including Singer.

[73] "The capacity for suffering and enjoyment *is a prerequisite for having interests at all.* A stone does not have interests because it cannot suffer ... A mouse for example does have an interest ... because it will suffer ..." Singer, *op. cit.* p. 7-8.

[74] Similar recognition of animal's legal right and obligation is found in some accounts of animal trial in the sixteenth century Europe where colony of we evils invaded some vineyards in France or also against beetles or others for similar legal offenses. The case of the insects was duly defended by lawyers appointed by the authority to which the villagers made petition for legal action against these tiny insects. After adequate hearing in some cases the culprits were punished (mostly banished from that place) if found guilty and if not, then it was victory for the insects after the Judge says that they have the same right to consume plant as humans. Luc Ferry, *The New Ecological Order*, trans. Carol Volk.

tures to their biosphere without which it could not have survived.

The ascription of right only to men first, and then to animals is associated with the notion of freedom vis a vis nature. That means natural objects are not free, therefore not legal subjects. Rights assume freedom. More freedom means more rights. If this is a right view then destruction of forest, river, mountain, atmosphere cannot be legally (and morally) wrong because these objects are not legal subjects, nor free and therefore these are not somebody's property, indiscriminate exploitation or misuse can never be wrong.

It is clear that this kind of logical consequence from either anthropocentric or sentient-centric ecology will not be acceptable to anyone including the ecologists themselves mentioned above. The oddity of such consequence will be clear if we take the example of a hypothetical society where the exploitation of nature is rampant, but it is not undertaken by the capitalist consumerist big powers, but by the ordinary people. Moreover the benefit and the yield goes ideally to the people who are closest to the nature—marginal farmers and the forest dwellers only, changing their life and bringing definite improvements in their life styles including the women's condition, their livestock, and also the wild life. In that kind of case the leftist, the feminists, the animal right activists would all be happy including the moderate liberals, the latter because the living standard and quality of life has definitely been improved.

Undoubtedly this is a case where all the standard arguments based on contingent state of affairs are not applicable but even then it is not easy to say that this type of exploitation/devastation of nature is justifiable. And to establish that we need a more comprehensive concept of freedom and right which is not doctrinaire; on the contrary quite satisfactorily applicable and consistent with the basic principles of ecology that we mentioned in section 1 namely maximum diversity and least interference. To do that I shall refer to two cases of insensate objects and controversy regarding their rights. The first case is an extraordinary trial[75] similar to the medieval trials we mentioned. On the basis of this trial the question arises immediately that "we give legal rights to forests, oceans, rivers and other so-called natural objects in the environment - indeed to the natural environment as a whole."[76]

---

[75] The legal suit was filed by the powerful eco-association, Sierra Club alleging that the 'development' permits to Walt Disney Enterprise (a giant corporation) by the U. S. Forest service for the construction of hotel, amusement park, etc. in a wild valley, Mineral King, situated in the Sierra Nevada will destroy the aesthetic and natural equilibrium of Mineral King. The suit was rejected on the plea that Sierra club has no interests' in Mineral king and American law in principle protects only interests' and not abstract value.

[76] These words are from C. D. Stone, in an article "Should Trees have Standing? Towards Legal Rights for Natural Objects' in *Southern California Law Reviews* , 1972. He writes in the preface to his book (reprint of the article). 'Perhaps the injury to the Sierra Club is tenuous, but the injury to Mineral King - the park "Jural" person - the way corporations

The other case is an article entitled 'Duties concerning Islands' by Mary Midgley in *Environmental Policy:* A collection of Readings, University Park and London: Pennsylvania State University Press, 1983. In this essay the author (a philosopher with several works on ecology) proposes a rewriting of *Robinson Crusoe* where he is set upon devastating his island in a delirious mood. Robinson's anecdote, according to the author demonstrates the existence of duties toward entities other than men. And not just towards islands, there exist moral obligation toward "the dead, posterity, children, the insane, embryos, human ... sentient and non-sentient animals, plants of all kinds ... ecosystems ... the biosphere, oneself and God!" I am not so sure about Midgley's sweeping effort to put everything in heaven and earth within the range of our moral obligation, but it at least shows the way ecological theories should go, we cannot stop halfway if we consider the symbiotic relation in which everything in nature is interactive. We cannot take a part there and ascribe it freedom without affecting the rest. If there is an obligation to man and other sentient beings then we have similar obligation to the rest of it. This deep ecological insight is necessary to understand the original natural environment man was part of. The social contractual understanding of man and nature is responsible for treating man as something different. The deep ecological principle is based on the two main principles of diversity and least interference, which are consistent with the democratic norm of equalization of right. It is not an abstract philosophical doctrine because many non-industrial traditions are based on it as indicated earlier in the discussion. Tolerance or "live and let live" is the basic principle here. If we consider the meaning of the word "eco" which came from "Aeiko" that means "home" or natural habitat, our principle would be more meaningful, home is not just people but the entire surrounding including animate and inanimate objects. Ecology thus should be 'study of home or natural habitat' as a whole. This theme also constitutes the ideological foundation of the *Geen peace* movement which believes that 'Humanist value systems must be replaced by supra humanistic values that bring all plant and animal life into the sphere of legal and ethical considerations ... force will eventually have to be brought to bear against those who would continue to desecrate the environment." Ignoring the coercive part, the warning echoes the basic environment principle we should follow today.

Popper in his anxiety to safeguard the world against too much legislation which needs coercion ignores the positive points of ecological movements which are of grave concern to any one. The spectre of Nazism led him to mistrust strong legislation against ecological devastation which might lead to

---

are "persons" - the notion of nature having rights would have made a significant operational difference.' Although the case was lost eventually by one vote Stone's argument nevertheless convinced three Judges to vote in favour of Sierra club.

possible loss of individual freedom. He realized the seriousness of the situation, but unlike most eco-philosophers think that there are built-in corrective measures within science and technological methods and by simply condemning modern science we cannot solve the problems. This position requires that the problems can be identified and remedied from within the scientific tradition. The principles of rational ecology as proposed here can give a satisfactory account of an environmental policy that is in tune with the critical rationalist position of freedom and rationality and is neither regressive nor a champion of mindless desecration.

# BIBLIOGRAPHY


Anderson, Terry L. and Leal et. al., 'Freedom and the Environment': Reply to Critics, *Critical Review,* 8 (3), 1994.

Archard, David, 'The Freedom Not to be Free. The case of the slavery contract in J.S. Mill's *On Liberty*', *Philosophical Quarterly*, 40 (161), 1990.

Arenson , R. J., 'Mill vs. Paternalism', *Ethics,* Vol. 90. no. 4, July 1984.

Arthur, J. 'Freedom of Speech', in *Philosophical Books*, 38 (4), 1997.

Augustino, D. and Jarvie, I. C. (eds.), *Freedom and Rationality,* B.S.P.S. 117, Kluwer, 1989.

Bahro, R., *Socialism and Survival*, Heretic Books, London, 1984.

Basu, P., 'Reappraisal of the Civilizational Question: Why the Scientific Revolution did not take place in China or India', *Sandhan* Volume II, No. 2, July-December, New Delhi, 2002.

Bhaskar, Roy., *Philosophy and the Idea of Freedom*, Blackwell, Oxford, 1991.

Bhaskar, Roy., *The Possibility Of Naturalism*, Harvester Wheatsheaf, London, 1989.

Bhatt, P.R.., 'The Issues concerning Human Freedom', *Indian Philosophical Quarterly*, October 1997.

Blum, Alex., 'Tenets of Freedom', *Philosophical Inquiry*, 12, Winter-spring, 1990.

Bookchin, Murray, *The Emergence and Dissolution of Hierarchy*, Palo Alto, 1982.

Bookchin, Murray, *Towards an Ecological Society*, Montreal, 1980.

Brodbeck, May (ed.), *Readings in the Philosophy of Social Sciences*, MacMillan, New York/London, 1968.

Broner, Stephen., 'The Enlightenment and its Critics', *New Politics*, Spring 1995.

Carr, L. Craig., 'Kants Theory of Political Authority', *History of Political Thought*, 10 (4), Winter, 1989.

Chattopadhyay, Debiprasad, *What is Living and What is Dead in Indian Philosophy*, New Delhi, 1976.

Chaudhury, Mahasweta, 'Objective Knowledge and Psychologism', *Journal of Council of Philosophical Research*, January 1995.

Clark, Peter, 'Popper on Determinism', in *Karl Popper: Philosophy and Problems*, ed. A. O'Hear, Cambridge, 1995.

Clark, Randolph, 'Freedom and Determinism', *Philosophical Inquiry*, 12, Winter-spring, 1990.

Cooper, Neilsen, Patten (eds.), *New Essays on J.S. Mill and Utilitarianism*, (*Canadian Journal of Philosophy*, suppl. vol. V), 1979.

Cummings, Robert Neville, *The Cosmology of Freedom*, New Edition, SUNY Press, Albany, 1996.

Daniels, N. (ed.), *Reading Rawls*, Blackwell, Oxford, 1975.

*Danish Year Book of Philosophy*, 28, 1993.

Dickens, P., *Society and Nature*, Hervester Wheatsheaf, London, 1992.

Dixon, Keith, *Freedom and Equality*, Routledge, London, 1986.

Douglas, J. D., *Freedom and Tyranny*, Alfred A. Knopf, New York, 1970.

Dworkin, G., *Taking Rights Seriously*, Bristol, 1977.

Dowling, Keith., 'The Justification of Democracy', *South African Journal of Philosophy* 13 (4) November, 1994. 'Mill's Principle of Liberty', *Philosophy* 69, October, 1994.

Dray, W., *Laws and Explanations in History*, Greenwood Press, London, 1979.

Dryzeck, J. S., *Rational Ecology*, Blackwell, Oxford, 1987.

Earman, John., *A Primer on Determinism*, D. Reidel, Dordrecht, 1986.

Elwin, Verrier., *A Philosophy for NEFA*, Delhi, 1960.

Evans, Marks., 'Freedom in Modern Society: Rousseau's Challenge', *Inquiry,* 38 (3), Summer, 1995.

Ferry, L., *The New Ecological Order*, translated by Carol Volk, Univ. of Chicago Press, London, 1995.

Feyerabend, P. K., *Against Method*, New Left Books, 1975.

Feyerabend, P. K., *Farewell to Reason*, Verso, 1988.

Feyerabend, P. K., *Philosophical Papers*, vols. I and II, 1991, 1992.

Feyerabend, P. K., *Science in a Free Society*, 1978.

Feyrabend, P. K., *Three Dialogues on Knowledge*, 1991.

Feyerabend, P. K., 'The End of Epistemology', in J. Earman and others (ed.), *Philosophical Problems of the Internal and External World.*

Fox-Keller, E., 'Feminism and Science', *Sex and Scientific Enquiry,* edited by S. Harding and J. F. O'Barr, Chicago,1989.

Fox-Keller, E. (ed.), *Reflections on Gender and Science*, Yale, 1985.

Fox, Warwick, *Towards a Transpersonal Ecology*, Boston, London, Sharubala, 1990.

Fuller, Steve, *Philosophy, Rhetoric and the End of Knowledge,* Madison University of Wisconsin Press, 1993.

Gadgil and Prasad, 'Forest Management and Forest Policy in India', *Social Action,* vol. 27, No.2, 1983.

Ginet, Carl., 'Freedom, Responsibility and Agency', *Journal of Ethics* 1 (1), 1997.

Goodman, N., *Fact Fiction and Forecast*, Bobs Merrill, Indianapolis, 1953.

Gordon, Jill., 'J. S. Mill and the Market Place of Ideas', 1997. *Social Theory and Practice,* 23 (2), Summer, 1997.

Gross, P. H., *Birth, Death and Migration in the Himalayas*, Delhi, 1982.

Paul Gross., Norman Levitt and Martin Lewis, *The Flight from Reason and Science,* New York, 1996.

Guha, Ram Chandra (ed.), *Social Ecology*, Oxford University Press, Delhi, 1994.

Guha Ram Chandra., *The Unquiet Woods, Ecological Change and Peasant Resistance in the Himalayas,* Oxford University Press, Delhi, 1989.

Gay, Peter, *The Enlightenment: An Interpretation.* Vol. 2; *The Science of Freedom* New York, 1966.

Gellner, Earnest, *Reason and Culture*, Blackwell, 1992.

Griffin, James, 'Equality and Sen's Weak Equity Axiom', *Mind,* vol. XC (1981).

Habib, I and Raina, D., (eds.), *Situating the History of Science*, New Delhi, Oxford University Press, 1999.

Halbfass, Wilhelm., *Tradition and Reflection: Explorations in Indian Thought*, Suny Press, 1991.

Hacking, I., *The Taming of Chance*, New York, C.U.P., 1990.

Harding, S., *Whose Science? Whose Knowledge? Thinking from Women's Lives*. Ithaca, Cornell University Press, 1991.

Harding, S. and J. F. O'Barr (ed.), *Sex and Scientific Inquiry*, Chicago, 1987.

Harding, S., *Is Science Multicultural?* Bloomington, Indiana, 1998.

Hart, Robert A. De. J., *Ecosociety*, Natraj Publishers, Dehra Dun, 1984.

Hayek, F., *The Fatal Conceit*, Routledge, London, 1988.

Hayek, F., *The Counter Revolution of Science,* Glencoe, Illinois, 1952.

Hollis, Martin., 'Market Equality and Social Freedom', *Journal of Applied Philosophy*, 7 (1), March 1990.

Hook, Sydney (ed.), *Determinism and Freedom*, London, MacMillan, 1958.

Honderich, Ted, *Three Essays on Political Violence*, Oxford, Blackwell, 1976.

Horz, Herbert., 'Ecology as a Challenge to Philosophy', Munver ed., *Beyond Reason.*

Horvath, Charles M., 'The Social Equation: Freedom and its Limits', *Business Ethics, Quarterly*, 5 (2), April 1995.

Irzik, Gurol, 'What is good for Science is not always Good for Society', *British Journal for Philosophy of Science*, 36, 1985.

Jarvie, Ian, *The Republic of Science* , Rodopi, Amsterdam, 2000.

Jonas, H., *The Imperative and Responsibility*, London, 1979.

Kalokerinou, H., *Freedom and Choice*, R. M. Hare and J. P. Sartre edited.

Kekes, J., 'Reason and Practice', in Munver edited. *Beyond Reason.*

Koertge, Noretta (ed.), *A House Built on Sand; Exposing Postmodern Myths about Science*, NY, Oxford University Press, 1998.

Koertge,Noretta, 'Ideology Science and Free Society', in Munver edited. *Beyond Reason.*

Korab-Karpowicz, Wlodzimierz, J., 'Freedom and Tradition', *Dialogue* (2) 2, 1992.

Lande, Alfred, *Foundations of Quantum Theory*, London, 1955.

Lande, Alfred, *Scientific Papers Presented to Max Born*, London, 1953.

Lchrer, K., 'Freedom, Preference and Autonomy', *Journal of Ethics* 1 (1), 1997.

Lemons, J., 'Cooperation and Stability as a Basis for Environmental Ethics', *Environmental Ethics*, 3, pp. 219-33.

Leopold, A., *A Sand County Almanac*, Oxford, 1949.

Longino, H., *Science as Social Knowledge,* Mannheim, K., *Diagnosis of Our Time: Wartime Essays of a Sociologist*, Kegan Paul, 1956.

Marcuse, H., *One Dimensional Man, Routledge* & Kegan Paul, London, 1964.

Marx, K., *The Capital*, London, 1949.

Mill, J. S., *Civilization, Dissertations and Discussions Vol. I and II.*

Mill, J. S., *Dissertations and Discussions*, Review of De Tocqueville's *Democracy in America.*

Mill, J. S., *On Liberty*, R. B. McCallum (ed.), Oxford, 1946.

Mill, J. S., *System of Logic*, Longmans, 1956.

Mill, J. S., *The Subjection of Women*, M.I.T. Press, Cambridge, Mass and London, 1970.

Miller, D., *Critical Rationalism: A Restatement and Defense*, Chicago and La Salle, Open Court, 1994.

Miller, D., 'Propensities and Indeterminism', in *Karl Popper: Philosophy and Problems*, ed. A. O'Hear, Cambridge, 1996.

Milne, D., 'A Note on Popper, Propensities and the Two Slit Experiment', *British Journal of Philosophy of Science*, 36, 1987, pp. 66-70.

Minek, J. (ed.), 'The Problem of Rationality in Science and Philosophy', *Boston Studies in the Philosophy of Science*, 160, 1995.

Montague, Richard, 'Deterministic Theories', in R. H. Thomson (ed.), *Formal Philosophy*, New Haven, Yale University Press, 1974.

Mukherjee, Radha Kamal., 'An Ecological Approach to Sociology: Eco-contribution to Sociology', *The Sociological Review*, Volume .XXII, Number 4, October 1930, reprinted in *Social Ecology*, edited by Ram Chandra Guha.

Munever, G. (ed.), *Beyond Reason, Essays in Honour of P.K. Feyerabend*, Kluwer, 1991.

Naess, Arne., 'Paul Feyerabend — A Green Hero', Munver edited, *Beyond Reason.*

Nandy, A, 'The Future University', *Seminar,* 425, January1995, New Delhi.

Needham, Joseph, *Science and Civilization in China*, vol. VI, 3, Cambridge University Press, 1996.

Nielsen, K., 'Impediments of Radical Egalitarianism', *American Philosophical Quarterly,* vol. 18, no. 2, April, 1981.

Nozick, R., *Anarchy, State and Utopia*, Oxford, 1974, 1980.

O'Hear, Anthony (ed.), *Karl Popper: Philosophy and Problems*, Cambridge University Press, 1995.

Osterfield , D., *Freedom, Society and the State,* University Press of America, 1983.

Popper, K., *Quantum Theory and the Schism in Physics*, Rowman and Littlefield, New Jersey, 1982.

Popper, K., *Realism and the Aim of Science*, Rowman and Littlefield, New Jersey, 1982.

Popper, K., *The Myth of the Framework*, M.A. Notturno (ed.), Routledge, 1994.

Popper, K., *In Search Of A Better World*, Routledge, 1994, 1998.

Popper, K., *All Life is Problem Solving*, Routledge,, 1999.

Popper, K., *The Open Society and Its Enemies*, vols. I and II, Routledge, 1995.

Popper, K., *The Open Universe*, Rowman and Littlefield, New Jersey, 1982.

Popper, K., *Conjectures and Refutations*, Routledge 1963

Popper, K., *The Poverty of Historicism*, Routledge and Kegan Paul, London, 1961 (3rd edition).

Popper, K., *The World of Propensity*, Thoemmes, Bristol, 1990.

Popper, K., *Objective Knowledge*, Clarendon Press, Oxford, 1974.

Ralston, J., 'Values in Nature', *Environmental Ethics* 3, pp. 113-128.

Rawls, J., *Political Liberalism*. Columbia University Press, New York, 1993.

Rawls, J., *Theory of Justice*, Oxford University Press, 1971.

Redclift, M., *Development and the Environmental Crisis, Red or Green Alternatives?* Methuen, 1984.

Runciman, W. G., *Relative Deprivation and Social Justice*, Routledge and Kegan Paul, London, 1966.

Ryan, Alan, *J. S. Mill*, MacMillan, 1987.

Ryan, Alan, *The Idea of Freedom*, Oxford University Press, 1979.

Saari, David. J., *Too Much Liberty? Perspective on Freedom and the American Dream*, New York, Praegon, 1995.

Sagoff, M., 'On Preserving the Natural Environment', *Yale Law Journal*, 84, 1974.

Schultz, F. M., 'Three Hues of Freedom', *Philosophy of Education*, ed. M.S. Katz, Urbana, 1995.

Seal, B. N., *Positive Sciences of the Ancient Hindus*, Calcutta, 1956.

Sen, A. K., *On Economic Inequality*, Clarendon Press, Oxford, 1973.

Sen, A. K, 'Rawls vs. Bentham: AnAxiomatic Examinations of the Pure Distributive Problem', first published in *Theory and Decisions* IV, 1974, and also in N. Daniels (ed.), *Reading Rawls*.

Sen, A. K., 'Justice: Means vs. Freedom', *Philosophy and Public Affairs*, 19 (2), Spring, 1990.

Sen, A. K., 'Wellbeing, Agency and Freedom', *Journal of Philosophy*, 82, 1985

Sen, A. K., 'East and West: The Reach of Reason', *New York Review of Books* July 20, 2000.

Sen, A. K., *Rationality and Freedom*, Oxford University Press, New Delhi, 2002.

Serres, M., *Le Contract Naturel*, Paris, 1990.

Shaw, P. D., 'Popper, Historicism and the Remaking of Society', *Philosophy of the Social Sciences* I, pp. 299-308.

Singer, Peter, 'Ethics and the Limits of Scientific Freedom', *Monist*, 79 (2), 1996.

Singer, P., *Animal Liberation*, 2nd ed., New York, Random House, 1990.

Sokal, Alan and Jean Bricmont, *Fashionable Nonsense: Post-modern Intellectual's Abuse of Science*, 1998.

Soles, D. E., 'On An Alleged Tension in Mill', *SW Phil. Review*, 13 (2), 1997.

Skorupsky, J., *The Cambridge Companion to Mill*, New York, C.U.P., 1998.

Stone C. D., 'Should Trees Have Standing? Towards Legal Rights for Objects', *Southern California Law Review*, 45, 1972.

Spakovsky, I., *Freedom, Determinism and Indeterminism,* Martinus Nijjhof, The Hague, 1963.

Stove, D., 'The Subjection of J.S. Mill', *Philosophy*, January, 1993.

Suchting, W.A., 'Marx, Popper and Historicism', *Inquiry* 15, 1971.

Taylor Charles., 'What is Wrong with Negative Liberty?, in A. Ryan (ed.), *The idea of Freedom.*

Tribe, L. H., 'Ways Not to Think About Plastic Trees: New Foundations For Environmental Law', *Yale Law Journal*, 83, 1974.

Urbach, Peter, 'Is Any of Popper's Arguments Against Historicism Valid?', *British Journal of Philosophy of Science*, 29, 1978.

Van Inwagen, P., *An Essay on Free Will,* Oxford, 1983.

Waldron, J., *Liberal Rights: Collected Papers*, 1981-91, New York, 1993.

Walker, M. A. (ed.), *Freedom, Democracy and Economic Welfare*, The Fraser Inst., Vancuover, 1988.

Watkins, J.W.N., 'Freedom and Predictability, an amendment to Mackay', *British Journal of Philosophy of Science*, 22nd August, 1971.

Watkins, J.W.N., 'Methodological Individualism and Social Tendencies', in May Brodbeck (ed.), *Readings in the Philosophy of Social Sciences*, 1968.

Watkins, J.W.N., 'Three Views Concerning Human Freedom', in R.S. Peters (ed.) *Nature and Conduct*, McMillan, 1975.

Watkins, J.W.N., *Human Freedom After Darwin,* Open Court, Chicago and La Salle, 1999.

Weber, E., *Peasants into Frenchmen*, Stanford, 1976.

Westacott, E., 'Teaching Mill's *On Liberty*', *Teaching Philosophy*, 20 (3), Summer, 1997.

Whelan, R., *Mounting Greenery*, London, 1989.

Whelan, R., *The Limits of International Co-operation*, London, 1990.

Winch, P., *Idea of Social Science and its Relation to Philosophy*, Routledge, London, 1990.

## NAME INDEX

SUBJECT INDEX

www.ingramcontent.com/pod-product-compliance
Lightning Source LLC
LaVergne TN
LVHW011133300926
849160LV00008B/629